CREATIVE GOVERNANCE

Creative Governance

Opportunities for Fisheries in Europe

Edited by

JAN KOOIMAN, MARTIJN VAN VLIET, SVEIN JENTOFT

Routledge
Taylor & Francis Group

LONDON AND NEW YORK

First published 1999 by Ashgate Publishing

Reissued 2018 by Routledge
2 Park Square, Milton Park, Abingdon, Oxon, OX14 4RN
711 Third Avenue, New York, NY 10017

Routledge is an imprint of the Taylor & Francis Group, an informa business

A Library of Congress record exists under LC control number: 98045730

ISBN 13: 978-1-138-61347-8 (hbk)
ISBN 13: 978-1-138-61349-2 (pbk)
ISBN 13: 978-0-429-46376-1 (ebk)

Contents

List of Contributors

Wim Dubbink
Informatie-en KennisCentrum Natuurbeheer
Ministerie van Landbouw, Natuurbeheer en Visserij
Marijkeweg 24, P.O. Box 30, 6700 AA Wageningen, The Netherlands
E-Mail: w.dubbink@ikcn.agro.nl

Peter Friis
North Atlantic Research Studies Centre
Roskilde University
P.O. Box 260, DK 4000, Roskilde, Denmark
E-Mail: paf@geol.ruc.dk

Rob van Ginkel
Department of Anthropology
University of Amsterdam
Oudezijds Achterburgwal 185, 1012 DK Amsterdam, The Netherlands
E-Mail: vanginkel@pscw.uva.nl

Svein Jentoft
Faculty of Social Science
University of Tromsø
N-9037 Tromsø, Norway
E-Mail: svein@isv.uit.no

Jan Kooiman
Prinseneiland 50-52 hs
1013 LR Amsterdam, The Netherlands
E-Mail: jkooiman@xs4all.nl

Jacqueline McGlade
Plymouth Marine Lab
Prospect Place, West Hoe
Plymouth PL1 3DH, United Kingdom

Carlos Iglesias Malvido
Departamento de Economica Aplicada
Universidade de Vigo
Aptdo.874-Vigo, Galicia, Spain
E-Mail: ciglesia@uvigo.es

Kaija I. Metuzals
Plymouth Marine Lab
Prospect Place, West Hoe
Plymouth PL1 3DH, United Kingdom
E-Mail: kime@wpo.nerc.ac.uk

Jeremy Phillipson
Geography Department
University of Hull
Hull, HU6 7RX, United Kingdom
E-Mail: j.phillipson@geo.hull.ac.uk

Jan Willem van der Schans
Hudsonstraat 27b
3025 CB Rotterdam
Tel: 31 (10) 476 5210

Juan-Luis Suarez de Vivero
Dept. de Geografia Humana
University of Seville
Maria de Padilla s/n
41004 Sevilla, Spain
E-Mail: vivero@cica.es

David Symes
Geography Department
University of Hull
Hull, HU6 7RX, United Kingdom

Jean-Paul Troadec
Menez Perroz
29880 – Plouguerneau, France
E-Mail: jean-paul.troadec@wanadoo.fr

Nynke Venema
E-Mail: nvenema@compuserve.com

Martijn van Vliet
Faculty of Business Administration, Dept. of Public Management
Erasmus University Rotterdam
P.O. Box 1738, 3000 DR Rotterdam, The Netherlands
E-Mail: l.vliet@fac.fbk.eur.nl

The European Science Foundation (ESF) acts as a catalyst for the development of science by bringing together leading scientists and funding agencies to debate, plan and implement pan-European scientific and science policy initiatives.

ESF is the European association of more than 60 major national funding agencies devoted to basic scientific research in over 20 countries. It represents all scientific disciplines: physical and engineering sciences, life and environmental sciences, medical sciences, humanities and social sciences. The Foundation assists its Member Organisations in two main ways: by bringing scientists together in its scientific programmes, networks, exploratory workshops and European research conferences, to work on topics of common concern; and through the joint study of issues of strategic importance in European science policy.

It maintains close relations with other scientific institutions within and outside Europe. By its activities, the ESF adds value by co-operation and co-ordination across national frontiers and endeavours, offers expert scientific advice on strategic issues, and provides the European forum for fundamental science.

This volume arises from the work of the ESF Scientific Programme on Tackling Environmental Resource Management (TERM).

Further information on ESF activities can be obtained from:

European Science Foundation
1 Quai Lezay-Marnésia
F-67080 Strasbourg Cedex
Website: http://www.esf.org

Preface

In Autumn 1995, the European Science Foundation launched a scientific programme on environmental research in the social sciences by the name of TERM: Tackling Environmental Resource Management. TERM is intended as a forum for teams encompassing collaborative research programs at the European level. In the context of this TERM programme, four major research themes have been identified:

➤ the comparative dynamics of consumption and production processes;
➤ policy-oriented learning and decision-making: environmental management and policy instruments under uncertainty;
➤ forms of international environmental co-operation and their development;
➤ perception, communication and the social representation of environmental change.

This book is the result of a series of workshops organised within the context of the TERM program. This study on fisheries, problems and opportunities in governing fisheries in Europe will make a contribution to these themes. The research projects on which it is based and which are reported in Part II of the book had, for the greater part, been concluded or were up and running when the TERM program started. The empirical research reported should therefore be seen as a secondary analysis of already existing material; the other, more conceptual Chapters have especially been written for this volume. Together they enable us to contribute to the indicated TERM themes in varying ways.

We express our gratitude to the European Science Foundation, which made the organisation of the work for this book possible, and wish to acknowledge that three of the case studies reported in Part II of the book were financially supported by the European Commission.

We thank the Rotterdam School of Management's Research Institute ERASM (Erasmus University, Rotterdam) and the Faculty of Social Sciences (University of Tromsø) for their financial support for post-editing. Finally, we express our gratitude to Jeroen Warner, who turned contributions by non-native authors into accessible English, and kindly took care of the lay-out.

Amsterdam, Rotterdam, Tromsø:
Jan Kooiman, Martijn van Vliet, Svein Jentoft

PART I
SETTING THE STAGE

1 Rethinking the Governance of Fisheries

JAN KOOIMAN

Character and Objective of the Book

Diversity, dynamics and complexity characterise the world we live in. This also applies to the subject we are dealing with in this book: fisheries and its governance in the European context. Since fisheries is a diverse, dynamic and complex social and economic sector, it is an instance of broader issues in the way natural resources are managed.

If we take this diversity, dynamics and complexity of the natural and the social world as our most general assumption, the first question that comes to mind is whether these three characteristics are sufficiently taken into account. The answer to this question is decidedly mixed. It is easy to find references on its complexity, diversity and, to a lesser degree, on its dynamics. If one probes deeper into what is offered in the way of analysis, problem sketches and solutions, these turn out to be more often characterised by simplicity, uniformity and a static view than one would have expected and hoped for.

We think this oversimplification of models and theories in use, the uniformity in thinking about causes and effects and the lack of sensitivity for the dynamics of developments in fisheries is unsatisfactory. The real world does not look like the simplified, uniform and static picture these theories paint. Nevertheless, interest group representatives, administrators, journalists, politicians as well as many scientists use them, for lack of alternatives or because of ideological or other preferences. There is not much we can do about this lack of appropriate perspectives: creators and users of knowledge are free to use whatever they think best for their particular purposes. But at least we can try to broaden the body of knowledge in this area. We believe that we have something to offer: a new perspective on fisheries in its social, economic and natural resource aspects. We shall develop a perspective of fisheries and their management as a diverse, complex and dynamic phenomenon. Thus we hope to add to the debate on what constitutes the good governance of this industry.

Fish as a resource is in great danger. There are serious concerns about the state of seas and oceans, about the livelihoods and way of life of many thousands of fishermen, their families, the communities they live in and all those other people depending on the resource. We shall take these concerns to heart. In the work accounted for in this book we have come up with suggestions - some of a more conceptual nature, some with a more empirical character - that could lead to a more fruitful approach in which new solutions are 'fished for'. It is our belief that such new opportunities for fisheries and in particular the way fisheries are governed, cannot be found in the dominant paradigms. That is why we think a new perspective on governance of fisheries is needed, on the basis of which new opportunities can be sought. In this study, our contribution revolves around two central terms: governance and opportunities. There is a need for creative thinking on governance, which means crossing boundaries between scientific fields of endeavour, countries, concerns and sectors.

Our basic contribution to the ongoing analyses of, and debates around, European fisheries is to 'open up' a number of issues that inform the present state of affairs in fisheries research and management. Three of these issues particularly stand out: crossing boundaries between dominant scholarly approaches; offering new perspectives on aspects of fisheries on the agenda; and establishing links between management issues.

➤ *Crossing boundaries between perspectives of fisheries as a scientific area.* Two disciplines dominate the scientific concern with fisheries: biology and economics. In both disciplines particular theoretical orientations play first fiddle. (Marine) biology focuses on particular species, much less on systems of species. In economics, attention is primarily focussed on market failures and property rights. In our opinion, the broadening of the scholarly scope has to take place at different levels of abstraction. There is also attention to fisheries issues in disciplines such as anthropology and sociology, but insights from these disciplines have great difficulty in gaining access to the policy arenas where the decisions on the future of fisheries in Europe are taken. It is not just a matter of simply adding other disciplinary interests. The broadening is about expectations of 'value added' to what is already undertaken.

➤ *Offering new perspectives on aspects of fisheries that get scholarly and political attention.* Fisheries and its problems is almost completely looked upon form the perspective of 'catch' and 'harvest'. This is, in a way, understandable, since there is broad agreement that over-fishing and depletion of fish-stocks is one of the main problems in fisheries. We basically agree with this point of view. However, because of the overwhelming focus on this aspect of fisheries, other aspects have not quite received the attention they need: the market and its (mutual) dependence on catch, local conditions and globalisation tendencies in fishing and fisheries, their geographic socio-economic interdependencies and developments, and, thirdly, the interrelations of fisheries on the institutional level in terms of international co-operation and connections with other institutional settings such as environmental concerns. They deserve this attention in their own right - fish is an important food product, and for some (remote) areas in Europe fishing is the only regular source of income; the tension between localisation and globalisation is growing, and fishing is a multi-dimensional, institutionalised part of the world. But looking at neglected aspects of fish and fishing and taking those aspects in a broader context than catch and harvest may also open up current perspectives on the main problems observed.

➤   *Establishing some new emphases in the management of fisheries.* When considering fisheries as a sector, one is struck by the dominance of an instrumental focus on management issues, and in particular on the management of fisheries in terms of the quantity of fish caught. From a historical perspective this is understandable, considering the dominant interest in scholarly as well as political/administrative domains. All energy is directed at one aspect of fisheries: stock control. This instrumental emphasis is phrased in the word 'management'.

We think this instrumental management approach should be broadened. The direction we chose is to make two connected steps. One is from management to governance and the second is from a problem and solution orientation to an opportunity-creating orientation. In our opinion the (broader) governance concept offers a number of advantages above the (narrower) concept of management. This applies in particular to fisheries, where management is so much identified with the application, administration and implementation of a set of instruments. And where a concept such as 'co'-management is seen primarily in an instrumental sense and not in the broader perspective of 'co-responsibility' or interactive governance.

To conceptualise our own approach to institutionalised aspects of fisheries we want to introduce the connected concepts of governance and opportunity in terms of 'governance of opportunities' and 'opportunities of governance'. We elaborate on this pair of concepts since we think the abundance of attention to the nature of 'the' problems and 'the' solutions in fisheries has not sufficiently reduced the amount of dissatisfaction with the current and expected state of affairs.

In sum: the three areas in which we would like to contribute to the debate on fisheries governance are:

> ➢   empirical research in fisheries on other aspects than the dominant ones;
> ➢   conceptual development on governance and opportunities for fisheries;
> ➢   developing some new perspectives that may help to implement these new opportunities.

We do not claim to oversee the whole scenery of fisheries theory, research and practice. We concentrate on selected aspects which we think are central in the combination of theory development, research on empirical situations and application in the practice of governance of European fisheries.

The Analytical Perspective

We call the theoretical perspective for this study *social-political governance.* Social-political governance is to choose an analytical and normative perspective on 'collective' problem-solving and opportunity-creation. Collective, not in the sense that the care and development of fisheries is seen as a public task (the 'state') or a responsibility of the private sector (the 'market'), nor of the 'third sector' (civil society) by themselves, but basically as a shared task. This is not to say that private and public actors cannot have specific responsibilities; of course they have. But the context in which tasks are defined or responsibilities are carried out have a specific 'co' element in them, rather than everybody going it alone. So public responsibilities are carried out with a view to 'private carrying capacities'; private tasks are carried out with a view to 'public carrying capacities' and some tasks are carried out in terms of specific 'co-arrangements', where the carrying capacities are defined as a common responsibility.

Our basic assumption and normative belief is that in modern societies, there is a growing need for 'co' tasks and responsibilities, in addition to the specific relevance and need for separate public and private tasks and responsibilities. In societal sectors such as fisheries, there is a growing need to look upon problem-solving or opportunity-creation in terms *of a 'mix' of self-governing, co-governing and hierarchical governing.* Every problem and every opportunity needs a specific mix of the three governing types. This study shows that to solve certain types of problems, or to create certain types of opportunities the broader institutional conditions in which these processes take place, cannot be left outside of the discussion on the social-political governance of fisheries in Europe. This institutionalisation is closely linked to the 'mix' of the three governing types, which is also an expression of the diversity, dynamics and complexity of fisheries as a social, cultural and economic phenomenon. In most prevailing policy and management models aimed at solving collective social-political problems, *diversity, dynamics and complexity* of social situations, conditions and developments are either ignored, or considered 'nasty' complications.

A second reason for placing these traits in the margin of the debate is that taking them seriously requires a multi-faceted approach and/or interdisciplinary co-operation. Such collaboration is the exception rather than the rule in the scholarly domain, fisheries research included.

To handle diversity, dynamics and complexity conceptually, the concept of *interaction* takes central stage. By introducing the concept of interaction we hope to make diversity, complexity and dynamics of modern societies conceptually accessible. Diversity refers to the actors, entities or agencies participating in interactions. The more diverse the participants, the more interesting and renewing the outcomes of the interactions. Complexity refers to the patterns of relations within interactions. Modern societies derive much of their strength from the refinement and overlap of the relations around social, technological and economic 'primary processes'. Dynamics refers to the tensions within interactions. These tensions have to do with the general characteristic of social (and natural) systems between forces aiming at renewal and change and forces which want to keep things as they are.

The primary governing task of social-political 'governors', either public, private or in combined public-private roles is basically to *solve social-political problems*, and to *create social-political opportunities* in situations or contexts that are diverse, dynamic and complex. They can do this by making explicit use of interactions as forms of governing, because diversity of interactions expresses itself primarily in actors defining a problem-situation; complexity can be expressed in the relations between the aspects of the interactions which are involved in the problem or opportunity definition. And the dynamics can be expressed in the location of tensions that may underlie the problems and opportunities as they are defined. There are thus specific methodologies connected with the way problems and opportunities can be tackled in diverse, complex and dynamic situations, issues or areas of social-political interaction. In this sense we term social-political problem-solving and/or opportunity creation as primary tasks of social-political 'governors': *first- order governing.*

However, the conditions under which day-to-day problem-solving and/or opportunity-creation in social-political situations takes place can also be problematised. As we call direct problem-solving and opportunity-creation first-order governing, designing the conditions under which first-order governing takes place can be specified as *second-order governing*. Problem-solving and opportunity-creation are shifted here, as it were, to a higher level. Analogous to problems and opportunities found in first-order governing, we discuss them in terms of *needs* and *capacities* at the second-order governing level.

To gain a better understanding of 'who governs what', we shall distinguish three types of governing interactions *self-governing, co-governing* and *hierarchical governing*. Self-governing interactions are the least organised, spontaneous and unruly forms of governing. The social realm and many economic interactions can be considered to be self-governing. Co-governing is an organised form of governing. Most professional fields and other sectors of society basically take the form of co-governing. Recently, co-governing networks are increasingly found in the market, and certainly in areas such as welfare, education and health. Hierarchical governing can be regarded as the most organised and formalised form of interaction. Most state or government policies, laws and regulations take the form of hierarchical governing interactions.

As indicated, most governance of social-political sectors, fisheries included, is a 'mix' of the three modes. But orders and modes of governing do not take effect in a void. We always have to ask the question: who governs the governors? Here we come to an important but often neglected aspect of governance, the 'governing of governing'. We call this the *meta* aspect of social-political governance.

There are two combined answers to liberal or social-democratic societies to this question:

> Governors are governed by norms, qualitative criteria, principles such as phrased in constitutions, declarations of rights etc.

> They are governed by the 'governed'. This is known as a 'strange loop', the fundamental tangling of hierarchical levels. In the final and most basic sense, the elected are governed by the electors; the governed govern themselves.

Both elements, alone or combined, are the essence of meta-governing in liberal or social-democratic societies or in their subsystems.

Equipped with these tools we attempt to analyse the *governing of European fisheries*, the major patterns of interactions within and around fisheries in terms of self-governing, co-governing and hierarchical governing, and the way fisheries are governed in terms of first and second-order governing. All this is taken to be within the context of certain forms of meta-governing, that is to say, broad social-political conditions that favour or hinder governing as an interactive process between governors and governed.

The above conceptual framework enables us to show that, in order to create opportunities for and within fisheries, it is necessary to look at the sector in terms of diverse, dynamic and complex patterns of activities and the governing of those activities. Opportunity-creation is not something haphazard that takes place in a void: it can be a conscious use or formation of patterns of interaction in which, depending on certain (meta)conditions, public and private actors operate; partly in isolation, partly in forms of co-operation to tackle the diversity, dynamics and complexity of the situation at hand.

Central Issue Statement

Our study has four building blocks:
> the TERM programme, which expresses concern about issues on the boundary between social-economics and environmental affairs;
> the goal we want to pursue in this book: new opportunities for governing fisheries (in Europe) are found in broadening up concerns and issues: not just in the areas of economics and biology, but other theoretical perspectives as well, with due regard for societal diversity, complexity and dynamics;
> an analytical framework, which will enable us to conceptualise research findings in terms of aspects of governing and governance;
> four research projects which provide empirical insights into fisheries and aquaculture in different areas of Europe (the UK, the Netherlands, Denmark and Spain).

These building-blocks lead us to the following objective:

The development of a perspective on interactive fisheries governance that creates new opportunities for fisheries in Europe.

We shall take the following points into consideration:
> by using the concepts of governing and governance we give special attention to public as well as private tasks/responsibilities in guarding and developing

fisheries;

➤ by emphasising the combination of environmental, social and economic aspects and concerns of fisheries, we translate and interpret empirical case studies into broader notions of the interrelations of these aspects and concerns;

➤ by stressing the governing of opportunity creation in European fisheries we stress that many of the issues raised can be phrased not only in terms of problems (to be solved) but also in terms of new perspectives on its diversity, complexity and dynamics which have received insufficient attention;

➤ by confronting insights from different scholarly disciplines we seek to overcome inherent limitations of each of the individual disciplines and to gain conceptual and theoretical insights hitherto not systematically explored;

➤ by focussing on a number of conceptual issues related to renewal of governance we seek to contribute to the debate on the renewal of the Common Fisheries Policy (CFP) of the European Union in 2002.

Outline of the Book

The above considerations also provide a framework for the outline of this book:
Chapter 2 discuss the existing state of the literature in terms of the major modes of governing as practised in European fisheries these days. Chapters 3 through 6 report on four research projects which together constitute the empirical part of the book. These case studies will be analysed along the lines of the governance framework used in Chapter 7.

The third part of the book (Chapters 8 through 11) investigates opportunity-creation in European fisheries by presenting four 'essays' about different types of opportunities. Chapters 8 and 9 tend to take a biological point of view, while Chapters 10 and 11 look into governing opportunities, and problems related to them. Part IV closes the book with two more conceptual Chapters, both dealing with important opportunity aspects: learning and action. These two Chapters also integrate insights from the preceding Chapters.

2 Evaluating Governance
State, Market and Participation Compared

MARTIJN VAN VLIET AND WIM DUBBINK

Introduction

This volume explores new opportunities for governing fisheries in Europe that take into account the complexity, dynamism and diversity of fisheries and fisheries management. Fresh and creative ideas often arise from a combination of new perspectives and concepts and new empirical material. This volume confronts the reader with the application of new perspectives on fisheries management to analyse systems of fisheries and aquaculture management in Europe. These perspectives do not arise in a vacuum. A parallel upward movement in perspectives depends on the way fisheries are managed and how they should be managed has accompanied the sharp increase in fisheries management efforts over the past three. New perspectives on managing fisheries, such as these based on Jan Kooiman's interactive governance approach, should be compared to other, more established views on managing fisheries. To define the arena, this Chapter will discuss and compare the three perspectives on governance that play a key role in the current debate.

Thereafter the challenge of governance in European fisheries is discussed and the three perspectives on governing fisheries analysed, each of which try to tackle the fisheries governance problem in its own way. These perspectives are not necessarily mutually exclusive. They highlight different aspects of the problem of fisheries management and, therefore, put forward different suggestions for reform. Next, the three perspectives are compared for commonalities and differences. The Chapter is concluded with a section on some critical issues in the current debate on the future of fishery management.

The Problem of Managing Fisheries[1]

The scientific literature on fisheries considers fisheries management to be an outstanding examplar of the problems of a sustainable use of renewable natural resources. The three basic and interrelated problems in fisheries are:

➢ the *biological* problem: the threat of over-exploiting the resource, so that the fertility of the fish stock is affected;

➢ the *economic* problem: the over-capitalisation and over-expansion of the fishing fleet;

➢ the *environmental* problem: the negative consequences of fisheries on the survival of sea wild life (sea mammals, birds) and its natural habitats.

Although scientists differ as to the causes of and solutions to these threats, they all agree upon the need for regulation of fisheries. Practical experience in managing fisheries affirms the correctness of this consensus. The insight that fish are prone to overfishing actually occurred to man a long time ago. Fisheries regulations go back at least as far as the Middle Ages. However, offshore fisheries were not regulated until a few decades ago. For a long time it was doubted whether the stocks at sea could be over-fished. Even at the beginning of the twentieth century, scientists as serious as Alfred Marshall were not sure:

> In river-fisheries, the extra returns to additional applications of capital and labour shows a rapid diminution. As to the sea, opinions differ. Its volume is vast, and fish are very prolific; and some think that a practically unlimited supply can be drawn from the sea by man without appreciably affecting the numbers that remain there; or in other words, that the law of diminishing returns scarcely applies at all to sea-fisheries: while others think that experience shows a falling off in the productiveness of those fisheries that have been vigorously worked, especially by steam trawlers. The question is an important one, for the future population of the world will be appreciably affected as regards both quantity and quality, by the available supply of fish. (Marshall, 1920, Book IV, Ch. III, para 7).

These days we are sure, but this knowledge has not brought us much relief. A new - and perhaps even more urgent - question has popped up: how can we regulate fisheries in an adequate manner; what is a good management regime?

The need to address this question is spurred by the fact that many of world's fisheries at the end of the twentieth century are in crisis (FAO, 1995). This crisis has two aspects. First, the crisis of stock conservation and over-expansion of the fishing fleet, as nicely depicted by Pearse:

Today, however, the regulatory regimes of major fishing nations are in turmoil. Viewed in historical perspective, the trouble appears as a crisis that was inevitable, sooner or later, given ever-expanding demands on the oceans' limited natural resources. It is a crisis marked by overexploitation and depressed stocks, continuing expansion of fishing power, advancing technologies for finding and catching fish, and competition among fishing nations and fishing groups over harvesting rights. (Pearse, 1994: 76-77)

The second aspect is a crisis of governance. Somehow, governments almost everywhere run into trouble while managing fisheries. Despite sometimes enormous effort, they rarely succeed in conserving the fish stocks. 'Rational' fisheries management seemed possible when 90% of commercial fish stocks were brought under a form of national jurisdiction as result of - or in anticipation of - UNCLOS III (Balton, 1996: 127). However, twenty years later, expectations for rational management are definitely more moderate. The 'cod that disappeared' from Canada's East Coast, but also the dangerous state of many fish stocks in the European waters call into question the effectiveness of the current management systems and the scientific perspectives these management systems are based on (Finlayson, 1994; Crean and Symes, 1996).

Particularly this second aspect is reflected at a theoretical level in the social sciences, especially in disciplines that analyse and try to improve public policy. These disciplines are found in the study of law, economics, political science, public administration and sociology. For want of a better word, we shall henceforth refer to 'policy sciences'. Many of the policy scientists try to cope with the phenomenon of 'ungovernability' that seems to haunt Western society since the 1970s (Offe, 1984). As modern societies become ever more diverse, dynamic and complex, and the role of government in society has grown continuously, governments find it ever more difficult to perform effectively, efficiently and legitimately (Kooiman and Van Vliet, 1995: 44).

The same governance problems occur in managing fisheries. Fisheries are difficult to manage because they are dynamic, diverse and complex. Economic, social, ecological and political interactions are such that stable, long-run balances are unlikely to occur. The abundance of fish stocks is not only influenced by fishing efforts but also by environmental factors such as (short and long-run) climatological changes and changes in food supply. In an increasingly internationalised market, market conditions for a particular kind of fish are influenced by supply and demand relations in the global fish market and influenced by price developments in competing products such as beef, pork and chicken. Fisheries are marked by a diversity of involved interests: there are small-scale and large-scale fishermen; community versus corporate interests; fishing and ecological interests. As a result, fisheries are complex: the density of the linkages and interactions between the actors and entities within fisheries and between fisheries and the environment is increasing. The outcome of these characteristics is a sector where:

➢ problems occur as the result of an interplay of various factors, some of which are clearly known, but others are easily overlooked;
➢ knowledge about causes and solutions is dispersed over many actors;
➢ uncertainty is the rule, not the exception.

Public authorities that try to intervene in fisheries management are faced with increasing problems carrying out the increased tasks and functions this involves. As a result, policy scientists are on the look-out for answers to these problems of governance and are evaluating and trying to develop new techniques, new instruments and new institutional arrangements that cope with these challenges.

In line with familiar classifications used in the policy sciences, this Chapter will focus on three perspectives of governance that dominate the theory and practice of fisheries management. These are:

➢ hierarchical governance;
➢ market governance;
➢ participatory governance.

Hierarchical governance is most regularly applied. In this perspective the government is held responsible for an adequate management of the fishery and uses its legal and administrative powers to enforce on the sector the rules and regulations it deems necessary. Public policies are often implemented this way. Fisheries are mostly regulated by legal and administrative measures that decide when, how and where fishing is allowed to take place.

Although (still) dominant, this perspective has come under severe criticism. In- and outside fisheries management studies, the hierarchical system is deemed outdated and inadequate for a complex modern world (Stone, 1975; Bardach and Kagan, 1983; Kooiman, 1993). Many scientists believe that the overtly familiar problems of compliance, implementation and control can be linked to the fact that government policy is overly organised. That is why two alternatives have come to the fore here.

The first alternative is *market governance*. Its advocates suggest that the government should make use of the market mechanism as much as possible by creating markets or market conditions (Schelling, 1983; Norton, 1984; Neher, ed. 1990; Hannesson, 1993; Pearse, 1994). In fisheries the ITQ system is the prime example of this model: by carving up the total allowable catch into private (property) rights that can be freely traded a market is created or imitated. Under these circumstances the profit orientation and cost-consciousness of individual fishermen will result in the most efficient way of harvesting the fish for society as a whole.

The second alternative is *participatory governance*. Advocates of increased participation in the process of governance hold that government should not try to do the job of regulating fisheries alone. The inherent 'complexity' of fishery management, due to the unpredictability of biological, economic and political circumstances, prevents one institution from having enough knowledge, legitimacy and power to regulate fisheries in an adequate way. Some responsibility for collective problem solving should be transferred to user-groups and/or other stakeholders, so that dispersed knowledge and governing capacities can be united (Dryzek, 1987; McCay and Acheson, 1987; Weale, 1992; Jentoft, 1995; McCay and Jentoft, 1996). The next section will describe and analyse the three perspectives.

Three Perspectives of Governance[2]

Introduction

In this section, each of the three perspectives is separately described and analysed. For each perspective its main characteristics, its reading of the fisheries governance problem in general, its explanation of the failure of current government policies of fisheries and its solutions or suggestions for reform are presented. We shall draw on representative works on recent fisheries management literature for each perspective. The hierarchical perspective is presented and defended in Mike Holden's book *The Common Fisheries Policy* (1994). The market perspective is represented by Pearse and Walters (1992) and Hannesson (1996). The participatory perspective is represented by McCay and Jentoft's recent article in the *Kölner Zeitschrift für Soziologie und Sozialpsychologie* and some chapters in Kevin Crean and David Symes' volume *Fisheries Management in Crisis* (1996).

Hierarchical Governance

Since discussions about governance so often start from a criticism of hierarchical governance, we shall start from this perspective. It draws a sharp distinction between society (including the market) and government. The latter is made almost solely responsible for the achievement of public objectives. To be more precise: the government is made responsible for the public objectives that are not provided by the market. In this view, the government should intervene on behalf of the public interest when the outcomes of unregulated social interactions are not in accordance with the perceived public interest. The government is held to be *legitimised* and *able* to regulate society within constitutional limits, in which the basic principles cherished by liberal-democratic societies are laid down.

The hierarchical perspective stresses that government intervention is legitimised when it is based upon *rule by law*. Just because in free and pluralist societies the proper role and tasks of government cannot be decided on for once and for all and are subject to ongoing debate, rule by law is considered a highly valued necessity because, in the end, the law and the legal system constitute the basic mechanism by which the danger of a potentially omnipotent government can be controlled. Hierarchical governance on basis of the law has four consequences that are considered advantageous in a modern, liberal-democratic society:

(1) Laws and regulations are made in a parliamentary-democratic procedure.
(2) The legitimacy of these rules and government intervention on the basis of these rules is guaranteed.
(3) The application of these rules is subject to independent juridical review, so that considerations as due process, impartial treatment and accountability are seen to.
(4) In principle clear distinctions are drawn between 'correct' and 'incorrect' behaviour and applied to everyone.

Ad (1): In a liberal democracy, government interference in private affairs is legitimised with an appeal to 'the public interest'. One of the most important functions of parliamentary law-making is the submission of laws and rules to a public and open discussion in which these laws and rules have to be defended in terms of the public interest. In this process social interests and stakeholders have the opportunity to participate in the discussion and to influence the decision-making process. Because of the very indefiniteness of the public interest, the parliamentary procedure safeguards that what is going to 'count as' the public interest is decided upon in a public and open debate.

Ad (2): Government intervention has to take place on the basis of prevailing laws because a government can only exercise its authority when it is explicitly allowed to do so. This enables the juridical review of government actions.

Ad (3): This examination and assessment has to take place by an independent judge. In case of a dispute on the implementation and application of laws and regulations in specific cases, subjects of these laws and regulations have the opportunity to bring their case before an independent judge so that values like equality before the law, legal security, impersonal appliance of the law, due process and accountability are guaranteed.

Ad (4): When it is laid down in laws and regulations what commands and norms are to be obeyed, a clear distinction between 'permissible' and 'impermissible' behaviour is created. This creates security for the subjects of these rules and creates for them the opportunity to adjust their decisions and actions; in the expectation that others (their neighbours, colleagues, competitors) are subjected to the same rules.

In sum, from a hierarchical perspective, the powers of government to regulate society are constrained by the obligation to legitimise its actions. The necessary legitimation is obtained by deriving her authority from laws that are agreed upon by the representatives of 'the people'. The guarantees within the process of law-making and law-appliance restrict the dangers of too big a concentration of power and an unjustified encroachment on the private sphere of society. However, when these guarantees are fulfilled, the government is legitimised and able (even obliged) to interfere in society.

Since the end of the 1960s, many policy scientists have analysed and explained regulatory and enforcement deficits and the implementation failures of government programmes (Pressman and Wildavsky, 1973; Mayntz, 1977). Often mentioned factors that are supposed to have caused the deficits and the failures of public policy are:

➤ Lack of knowledge, time and financial resources within the public service, especially on the part of enforcers;
➤ The power of social groups, especially business to influence public decision making or to resist government pressure;
➤ The complexity of the issues at stake, which encumbers or even prevents the design of adequate regulatory devices.

For students of fisheries management these causes sound familiar. In empirical studies on the deficits of fisheries management there are many references to the complex rules and regulations concerning such aspects as the kind of gears to be used, maze widths, motor capacities, and periods and areas in which fishing is permitted; to the lack of resources on the part of compliance agencies in the face of the difficult control and enforcement tasks flowing from the complexity of the rules; and, last but not least, to the influence fishermen's organisations seem to have on the public decision-making process.

Implementation and enforcement of fisheries rules and regulations is better considered a negotiation game that prevents public policy objectives being fully met. Downing and Hanf (1983: 318) mention three characteristics of implementation of environmental regulation that are also applicable to the implementation of fisheries rules: multi-actor interactive process, strategic behaviour with bargaining among key actors and interaction of interest-motivated actors under constraints, i.e. mutual dependence and limited resources. Because of these characteristics, the result of these actions is incomplete enforcement.

The hierarchical perspective acknowledges that European fisheries management is subject to implementation, enforcement and performance deficits. However this deficit is not considered a result of fundamental failures in the hierarchical mode. Deficits in European fisheries management are the result of an incomplete application of the Common Fisheries Policy (CFP). A lack of control and weak enforcement are the result of an insufficient specificity of the CFP's objectives, the absence of a unified power and the lack of determination ('guts') on the part of European politicians.

In his excellent and deep analysis of the making and implementation of the Common Fisheries Policy, the late Mike Holden (1994) shows himself to be a defender of the hierarchical position. He holds the opinion that the CFP wants more specific objectives (p. 222 ff.; 'maximising the economic benefits from its fish resources', p. 233) and that it is the responsibility of politicians to take the decisions that bring these objectives within reach (p. 232). His proposals for implementation of these objectives (Ch. 12: 239ff.) are centered on a strengthening role of the Commission in managing European fisheries.

In his opinion, the principal cause of CFP ineffectiveness is that the management responsibilities are split between the European Commission and the Member States. His argument is summarised as follows:

> One of the fundamental reasons for having the CFP is the recognition of the fact that managing the harvesting of a common resource requires agreement between its owners (...). However, under the present system, the management

of this resource is split. Management measures are adopted by the Community but their implementation is the responsibility of the Member States (...) but is this division of responsibilities compatible with effective management? (...)

(T)he majority of Member States do not take their responsibilities seriously and, in consequence, the whole policy is failing. *For management to be effective, responsibility cannot be divided. Management cannot be left to the Member States. Management must be the responsibility of the Commission; that is its institutional role.* (Holden, 1994: 245, emphasis added).

To solve the current crisis, Holden insists on applying a Commission-managed European licensing system to reduce fleet capacity (p. 245ff.).

Brief, Holden proposes a unified management system that he explicitly legitimises with references to the European 'Constitution', the Treaty of Rome, and the underlying principle of subsidiarity. In his view, the European Commission is capable of governing this system if the member states recognise the need for it and have the political will and courage to 'break the mould' and set aside the CFP's cornerstone, the relative stability principle.

Market Governance

Advocates of market-governance assume public problems to arise from 'market failure' (Baumol and Oates, 1975; Pearce and Turner, 1990). 'Market failure is said to occur when the price mechanism or the market system, the so-called invisible hand, fails to bring about a social optimum' (Tisdell, 1993: 7). Neo-classical economic theory assumes that in a perfectly operating market, the interactions between self-interested actors lead to the most efficient (that is, socially optimal) situation. However, if there is a market failure, individuals are induced to behave in ways that do not contribute to the social optimum; the incentive structure being such that private and public interests diverge.

If we accept the proposition, then, that 'market failure' is at the root of government failure, the continuing problems of hierarchical governance in fisheries and elsewhere can be eloquently explained. Hierarchically organised governments correct market failures by legal and administrative measures. These measures, however, do not resolve the market imperfection, and therefore do not diminish the divergence between private and public interest. As a consequence, governments have to put in an enormous effort to steer unwilling subjects.

According to advocates of market regulation, it would be much wiser for the government to try and change the incentive structure such that private and public interest are reconciled. The workings of the perfect market should be restored. When the market is restored, we draw on the *self-interest* of people in implementing public policy. The government will then have a much easier task. Thomas Schelling quite nicely sums up the ideas underlying market regulation:

> Activities that impinge on the environment are regarded (...) as activities that are, but need not be, external to the economic system - outside the market, unpriced, unowned, unmeasured, and not accounted for by those whose activity does the impinging. The impact is an economic 'output' that differs from other outputs in not being comprised within private transactions. (Advocates of market-based perspective) recognise that there are problems in bringing these activities within the realm of economic accountability, but usually expect the problems to be solvable and do not see why the activities should not be subject to economic principles. (Schelling, 1983: ix).

This line of reasoning is very common to fisheries management literature. The problems related to fisheries are often conceptualised as market failure (Gordon, 1954; Neher, 1988). Fishery economists who are in favour of market regulation, have indeed analysed the situation in fisheries precisely along these lines. Gordon and his successors set out 'to demonstrate that the "overfishing problem" has its roots in the economic organisation of the industry' (Gordon, 1954: 128). Because fishermen make use of a common-property resource: 'the fish in the sea are valueless to the fisherman, because there is no assurance that they will be there for him tomorrow if they are left behind today' (Gordon, 1954: 135).

The problem of fishery governance is considered the outcome of market failure resulting from the common-property character of fish stocks exploited by profit-maximising fishermen. A 'rational', rent-seeking fisherman is incapable and unwilling to forgo short-term profits in favour of long-term proceeds that should be the consequence of the preservation of a bigger stock in the sea. The common-property character of fish stocks and fishing grounds prevent unregulated fisheries to be efficient. The familiar economic theorising is summarised in the well-known Gordon-Schaefer curve that shows us that the point of equilibrium is reached in a situation where more fleet capacity than necessary catches less fish than possible (Gordon, 1954; Cunningham, 1985; Pearse, 1994, Hannesson, 1996).

As a consequence, the crisis of governability that currently harasses the fisheries can hardly be a surprise. Given the fact that public and private incentives diverge, it is a possibility to make and implement rules that directly determine who is allowed to fish what, where, when and how, so that the market failure is corrected. However, hierarchical governance has an automatic failure mechanism, in the sense that when a government is successful in regulating fisheries, the divergence between private and public interest increases, and the fishermen's financial incentive to cheat and circumvent regulations grows.

Following the assumptions underpinning the market-regulation model, restoring the market is the best answer to any market imperfection. This explains why individual transferable quotas (ITQs) are so popular with fisheries economists. ITQs can be looked upon as a first, but major step towards full individual property rights (Pearse, 1994). ITQs, therefore, help to restore the ideal workings of the market mechanism, thereby assailing the problems of governability at its root.

Advocates of market regulation suppose that ITQs, especially if they are further evolved towards real property rights, could solve the biological, economic and administrative problems of the fisheries to a considerable degree. The (threat of) depletion of the fish stocks - the biological problem -will be alleviated to the degree that ITQs enable individuals to act as private owners of a resource (who are supposed to have an interest in the conservation of the stock).

The economic problem, 'which is manifested in the waste of labor and capital in redundant catching capacity, excessive costs, (and generally) depressed incomes...' (Pearse, 1994: 7) will be softened inasfar as ITQs enable individuals to escape the prisoners-dilemma, and decide independently when and how to apply and organise the available 'production factors' (short-term efficiency increase). Furthermore, since the quotas are transferable, the market mechanism will stimulate that they are priced, thereby inducing a trend for quotas to be transferred from the least efficient to the most efficient fisherman (long-run efficiency increase).

The enormous involvement of modern governments in fisheries - the administrative problem - will also be reduced, but it is hard to say to what extent. Even a perfectly organised market could not do without some administrative effort. However, since the incentives to trespass, would become less irresistible, the effort would considerably diminish in comparison with the present situation.

All this is not to say that advocates of market regulation believe that the introduction of ITQs, would solve all the problems of the fisheries just by themselves.

Even quota licenses fall well short of complete or perfect property as defined by lawyers and economists, which requires that rights be exclusive, perpetual, divisible and transferable, and convey all the economic benefits to its holder, among other things. Most important, the holder of a quota does not have an exclusive right to a stock of fish; he shares it with other quota holders (Pearse, 1994: 86).

Pearse, a distinguished advocate of market governance, realistically qualifies the pretences of the ITQ systems, but does not qualify the market approach as such. It is fair to say that according to Pearse, the degree to which market instruments fail to solve the problems of the fisheries is a measure of their inability to close the gap between public and private interests so far.

Participatory Governance

Arguments in favour of participatory governance normally start by criticising hierarchical modes of governance. But whereas advocates of market regulation focus on difficulties at the *instrumental* level, do advocates of participation focus on the *organisational* level. The participatory perspective considers the fact that governance is mainly organised at the macro level of state bureaucracy a main cause for the problems surrounding public policies. The meso level, the level of civil and private organisations, is presently hardly made relevant for governance. The micro level of the individual (or the individual firm) is only passively involved, as the subject of government-regulations.

In hierarchical governance the public responsibility for managing the fishery is placed solely at the state level. According to advocates of participatory governance, the vacuum at the meso and micro level is an open invitation to implementation and compliance problems. Since we know that the government is far from triple perfect (having perfect knowledge, perfect judgement and perfect power all at the same time) we can be sure that the institutional neglect of the meso and micro levels by the government will cause serious trouble.

The solution that participation offers to the problem of ungovernability reflects the *organisational focus* of its proponents: to solve the problems of hierarchical governance, the structure of governance has to be changed. Central to the perspective is the idea that the meso and micro levels of society should participate more in governing, and significantly so. Citizens and 'stakeholders' should be more actively involved in the formulation and implementation of public policies. The term 'co-management' that is often associated with the participatory governance perspective directly refers to the organisational changes that need to be made. The government should not govern alone. Organisations and individuals at the neglected levels should take their share.

The participatory perspective implies a less legalistic approach of governing at the micro-level, leaving more discretion to individuals and firms to adapt their conduct to 'the spirit of public policy'. Advocates of participation do not consider this to be a naive point of view. On the contrary, they assume their approach to be built on the realistic idea that for one reason or another many regulations actually remain a dead letter, since they cannot be enforced at the micro level, at least not in the way the government meant them to function (Stone, 1975; Mayntz, 1978; Kagan and Bardach, 1983; Maus, 1986).

Moreover, the advocates of participatory governance make the less legalistic approach at the micro level dependent on more important changes at the meso level of society. Here an institutional structure needs to be built where industries, the government and representatives of citizen groups debate and negotiate on public policies. The emphasis placed on the importance of strengthening the meso level of governance is actually one of the fundamental differences between strategies directed to increased participation and strategies like 'deregulation'.

Proponents of participation do not believe that 'self-government' and less explicit legislation can have any meaningful impact in case the meso level remains neglected. In order to solve the complex collective problems, there is need for *deliberate* co-ordination in modern society. This sort of co-ordination cannot be provided for by (individuals acting on) the market.

There is some difference of opinion with regard to the lay-out of the governance structure at the meso level among advocates of co-management. Some advocates explicitly assume that all parties concerned should be involved (Dryzek, 1992; Weale, 1989). Others are vague on this point or believe that co-management is nothing but building up organisations in which government and representatives of industry negotiate about public policy (Glasbergen, 1989; Jentoft, 1995).

Advocates of participation place great stress on the need to re-arrange the organisational structure of governance. But it would be misunderstood if it were exclusively interpreted as an organisational face-lift. If proper institutional provisions are provided for, arrangements based on participation of resource holders and other stakeholders are contingent on the idea that deliberation and negotiation at the meso level can be more than a bargaining process in which the negotiating powers hold a pistol to each other's heads. The implementation of the participatory perspective therefore requires a change of attitude among participants. Both the regulator and the regulated have to operate on the basis of greater mutual trust. Otherwise, discretion could impossibly be given to business. This is why participatory governance takes ideas about the possibilities of 'communicative rationality' and 'social control' more seriously than hard-boiled realists would (see Dryzek, 1987).

Its advocates claim that participation has some great advantages over hierarchical regulation. Because it is less legalistic and involves the micro and meso levels of government, it is held to be much more able to adapt to complexity of governance in modern society. In addition, it can be assumed that participation will boost the legitimacy of government (Ostrom, 1990; Jentoft, 1995). This will both augment compliance to present rules and diminish the number of clumsy future rules, for individuals at the micro-level will be more involved.

All in all participation presumably leads to rules and measures that are better in substance, better complied with, and reduce governmental 'overload'.

Hierarchical, Market and Participatory Governance Compared

Now that the essentials have been laid bare, we can try to compare hierarchical, market and participatory governance. Making a comparison is a complex matter. We are not comparing empirical data but alternative *perspectives* for governing fisheries. Moreover, these perspectives focus on different aspects of governance. Advocates of the hierarchical perspective stress the *legality* of public policies. Advocates of market governance concentrate on the *instruments* the government should use. Advocates of participatory governance analyse the best way to arrange the *organisational structures* of society.

There does not have to be a radical opposition between these perspectives on every issue. Under specific circumstances they might come up with the same suggestions or they could be combined, but on other aspects they are not only taking different but incompatible angles on the fisheries management problem. Because this Chapter is intended to set the stage for a further exploration of the opportunities for new, more fruitful, ways of governing European fisheries, this section lays out the differences and similarities of these perspectives.

Common Denominators

Similarities between Hierarchical and Market Governance

Market governance shares three important ideas with the hierarchical perspective:

> ➢ both perspectives are based upon the belief that the responsibility for solving public problems resides with the government (whether or not it makes use of the market;
> ➢ both feel that private organisations and citizens have a limited responsibility for public problem-solving;
> ➢ both agree to the assumption that the government is potentially able to enforce itself on *civil society* including the economy.

Gordon (1954) and Hardin (1968) lay the foundations for the market and the hierarchical perspective. Both stress that users of unregulated common property face a situation they themselves cannot escape from. There is a need for a state or a 'unified directing power' (Gordon, 1954: 135) but they stress that the state has different institutional devices at hand.

Similarities between Hierarchical and Participatory Governance

The participatory perspective shares some other ideas with the hierarchical perspective. First, government actions should not only be assessed on efficiency and effectiveness criteria but also have to attain *normative* objectives. Second, the idea that legitimacy is an important aspect of government, which is related to the idea that the need for legitimacy binds governmental action to all kinds of procedural rules. Both perspectives stress that in liberal democracies legitimate and effective governance can only be effectuated with the consent of the people.

Similarities between Market and Participatory Governance

The market perspective and the participatory perspective agree that policy implementation by governmental organisations should be minimised as much as possible because it is at risk of overbureaucratisation and of a lack of efficiency. Both consider the heavy involvement of government agencies in executing the many rules and regulations to be a principal reason for the governability crisis in fisheries. A second common denominator is that both perspectives expect management systems that allow some degrees of freedom to users on how to run their businesses to be more easily accepted and complied with than more constraining management systems.

Differences

Partially the differences between the three perspectives are reflected in the above section. The perspectives differ on the characteristics they share with one perspective but not with the third one. In this section we concentrate on additional differences, mainly consisting of characteristics that are particular, i.e. not shared with another perspective.

Differences between Hierarchical and Market Governance

The hierarchical perspective sees bureaucratic mechanisms as resulting from the result of necessary procedures that safeguard (liberal-democratic) constitutional principles such as accountability, equality before the law and due process. These procedures determine the lawfulness and therefore the legitimacy of the applied regulations. Issues of legitimacy and normative considerations in general do not find an easy place in the market-regulation model. The 'legitimate' task of the government is the construction and safeguarding of property rights, so that the market can function optimally (i.e. efficiently).

Differences between Hierarchical and Participatory Governance

Both perspectives correspond in their interest in issues of legitimacy, but differ in their view on the basis for legitimate regulations. From the hierarchical perspective, the legitimacy of (specific) regulations and their application is dependent on their lawfulness: based upon laws that have the consent of 'the people' as represented in a (freely elected) parliament and are subject to judicial review (legitimacy by electoral representation). The participatory perspective suggests that in complex issues such as fisheries regulation, legitimacy is undermined by a low effectiveness. Therefore, legitimacy as result of direct participation and acceptance is stressed, to restore effectiveness and legitimacy of fishery regulations

Differences between Market and Participatory Governance

An important difference between market governance and participatory governance is that advocates of each model start from quite different assumptions on society and man. Advocates of market regulation conceptualise 'man' as an economic actor, and society (mainly) as a market; both in an empirical and a normative sense. The advocates of participation conceive of man as something more than an economic actor. People are also parents, friends and citizens, to name just a few important social roles. Society is also regarded as something more than a market. Modern society is conceived of as differentiated in various spheres, of which the market only constitutes one. That is why the market perspective has the tendency to describe social problems as 'market failures' and to suggest 'restoration of the market' as a solution, whereas the participatory perspective does not have these tendencies.

A related difference between market governance and participation is that advocates of the market solution usually evaluate the government only for its efficiency and effectiveness. They define (multiple) objectives and search for the most cost-effective way to achieve them. Advocates of participation believe that the government also has important *normative-procedural goals* to attain, since the government is in constant need of legitimacy.

Conclusion: the Central Issues in the Debate

The above section has compared the three governance perspectives. On the basis of this comparison the central issues relating to debate on new possibilities for governing (European) fisheries are discussed in this final section. We address the following six issues:

➢ What are the proper objectives of fisheries management?
➢ How far do the powers of government reach and how far are they allowed/legitimated to reach?
➢ What are the potentials and limitations of the market?
➢ What is the relation between public and private responsibility?
➢ What is the basis for co-operation between parties involved and which parties need to be involved in fisheries management?
➢ How do you organise the management structure?

1. The Objectives of Fisheries Management

The market perspective defines the most efficient exploitation of fish resources as the sole objective of fisheries management, under the condition that all possible social values carry a price tag. Both the hierarchical and the participatory perspective stress that normative and procedural objectives can be aimed at as well. In the opinion of some of the market regulation advocates, efficient exploitation of fisheries is often a necessity to survive international competition. It is stated, for example, that ITQ systems were introduced in Iceland because 'Icelanders simply cannot afford to run an inefficient fishing industry' (Arnason, 1996: 72). By contrast, the participatory perspective tends to stress social and moral concerns such as the viability of fishing communities or fishery dependent regions, and ecosystem concerns. These concerns are stressed not only in their own right but also since fishery management systems have to take these concerns and interests into account to be effective. Although the hierarchical perspective is in principle adjacent to the participation perspective on this point, it tends to define the objectives of fisheries management at the macro-level.

2. The Power of Government

The hierarchical and market perspectives both consider the government able and legitimised to regulate the fisheries sector, if in different ways. This is seriously questioned in the participatory perspective. The complexity of fisheries management is seen as the reason why fisheries cannot adequately be managed without involvement and support of the fishing sector itself. This is often viewed as the main legitimacy issue. 'Legitimacy', however, is more than acceptance or social support from the user groups/targets of regulation alone: it also involves the acceptance of other stake(holder)s such as environmental concerns and consumers. When participation in the form of user-group involvement is only meant as an instrument to increase the performance of public policies, then in a moral sense, participation is neutral in comparison to other policy instruments (cf. Goodin, 1986).

3. Potentials and Limitations of the Market

From the market perspective, the market is the most natural system in which men and women can interact. If the incentives are rightly constructed the market should be very effective in solving a broad range of public problems. The other perspectives suggest that 'getting the incentives right' is not as easy as is often suggested. Objectives can change over time, more and other knowledge could lead to changes in the perception on what is in the public benefit. The hierarchical and participatory perspective both stress that 'the market' is only one of the spheres in society, albeit an important one.

4. Public and Private Responsibility

The market perspective tries to construct a world in which there will no longer be any discrepancies between the private and the public interest by changing financial incentives. Such a society has no need for individual responsibility. People always act in the public interest. Proponents of the participatory perspective do not adhere to this ideal, nor do they assume it to be realistic. There will always be some tension between public and private interests. Therefore the participatory perspective stresses the necessity of delegating public responsibility to private parties (citizens as well as firms). The most important question in this perspective is how this delegation can be stimulated and how societal and market structures can be influenced such that private parties are able and willing to take public responsibility.

5. The Basis for Co-operation

From a market perspective, co-operation depends on and results from the refinement of property rights. Co-operation will then appear between the owners of quotas. The fishermen in this image co-operate to defend their private interest as individuals. They will act as if they are a 'sole owner' of the fish stock (Pearse, 1994).

Advocates of participatory governance do not make co-operation dependent on the extension of property rights; neither do they assume that the fishermen co-operate as if they were an association of owners, who are forced to co-operate to defend their private interest. They suggest that under certain conditions, fishermen will co-operate among themselves in the public interest, without a direct relation to their own interest.

6. The Management Structure

All this winds up to the question what a good management structure for fisheries should look like. The market governance perspective stresses the pecuniary incentives. The hierarchical governance perspective stresses the legal incentives. The participatory governance perspective stresses the normative incentives that influence the fisheries sector and the behaviour of fishermen. It seems that the design of an improved management structure should not be based upon a view that *a priori* gives one kind of wrongly structured type of incentives as the cause of fishery management problems primary attention. The question seems to be how legal, financial, and normative incentives on the micro, meso and macro levels interact and are balanced. The challenge is to develop a system of governance that creates room for co-ordinated collective action on the one hand, but enough flexibility for individuals on the other.

Notes

1   This section draws from Dubbink and Van Vliet (1997), pp. 177-179.
2   Part of this section draws from Dubbink and Van Vliet (1997), pp. 180-184.

PART II
LEARNING FROM EXPERIENCE

Learning From Experience
Introduction to Part II

The common denominator of the case studies presented here is that they all address complex areas of fisheries activities that deal with dynamic interactions between governments and private and civic partners representing a diversity of user groups.

In each case, complexity is expressed in the multiple use of common resources, in overlapping areas of social and economic activities and in transgressing tasks and responsibilities of involved actors and actor groups.

The common character in the dynamics presented in the cases is shown in the tensions emanating from changes taking place in the social and economic circumstances of fisheries and responded to through social-political adaptations to those changes - and the other way around. A common characteristic can be found in the almost unlimited variety of behaviours, opinions, interests, and cultural and organisational patterns.

The 'Mussels' case (Chapter 3) addresses issues in old aquaculture. Its central *problématique* is the modernisation of a traditional trade. The study shows definite but varying patterns of interaction between the public and private sector in the course of many centuries. The outcome, which shows shifting balances in social-political governance between public and private interests and forces, is important here. If this balance is effective at all, it is only effective in a certain context and for a certain period of time. There is no such thing as a long-run equilibrium between state and market in governing fisheries.

The central issue in the 'Co-Governing' case (Chapter 4) is how Fishermen's Organisations (and to a lesser extent Producer Organisations) can and should play a role in the governing of fisheries. This case study touches the core of the co-arrangements in the governance of fisheries, the role of public and private actors in their horizontal and vertical patterns of interactions. It also addresses to a considerable extent the diversity of these governing patterns, showing how in the three countries studied (UK, Spain and Denmark) institutional arrangements within the public and private sector have to be looked upon as decisive elements in the potential and actuality of those co-arrangements.

The 'Salmon' case (Chapter 5) addresses issues in today's aquaculture. The development of salmon farming, a capital-intensive activity that was sponsored by multinational companies from the start, is a manifestation of modernisation processes in itself. The *problématique* here is how this 'modern' activity interacts with the ecological, social and cultural environment it is placed in. It is a striking example of how a new line in fisheries is introduced where traditionally the dependency on more traditional forms of fisheries is great; and regions where the need to create new opportunities for work, constitutes a great social-political governing challenge.

The case study on 'Quality in the Chain' (Chapter 6) concerns itself with the possibilities of co-governing in fisheries in terms of the relation between its catch and its market side. In particular it looks into the possibilities of creating opportunities for the chain as a whole, by starting from the market side (demand) rather than the harvesting side (supply). What kinds of opportunities can be created for European fisheries if we look into the market potential in three countries: the Netherlands, the UK and Spain?

Finally, Chapter 7 brings some of the insights from these case studies together, analysing them with the help of concepts introduced in Chapter 1. In addition to greater coherence, it also offers the opportunity to see how the 'social-political' governance approach can be put to use in describing and analysing some of the main characteristics of European fisheries. By going through the concepts systematically, a clearer picture emerges of some of the strengths and weaknesses of governing these fisheries. It also enables us to phrase a number of conclusions that, taken together, can be seen as a (partial) answer to the condition of fisheries governance in Europe.

3 Capturing and Culturing the Commons

Public-Private Dynamics in the Dutch Oyster and Mussel Industry

ROB VAN GINKEL

But then came the explosion. The sleeping Giant, later known as the Consumer, awoke with a yell, and the cry for oysters and more oysters was heard across every civilised land (Clark, 1964: 43).

Nature is seen by humans through a screen of beliefs, knowledge, and sspurposes, and it is in terms of their images of nature, rather than of the actual structure of nature, that they act. Yet, it is upon nature itself that they do act, and it is nature itself that acts upon them, nurturing or destroying them (Rappaport, 1979: 97).

Introduction

It has almost become conventional wisdom that unlimited entry to the marine domain leads to over-fishing and that state intervention or privatisation provide solutions to this problem. Yet there is overriding empirical evidence that users of marine resources held as common property in many cases have developed rules and rights leading to their sustainable use. Access to common-pool resources is rarely open to all. This is not to deny that 'tragedies of the commons' exist. While overexploitation will occur under certain conditions, this is not necessarily attributable to the rapacious behaviour of fishermen. What is needed is a careful reconstruction of the factors that occasion resource abuse (see also Van Ginkel, 1996).

In recent discussions on how to tackle the problems of declining harvests and overexploitation of marine resources, fish and shellfish farming is often being hailed as a solution to the limitations nature poses. The underlying idea is that interventions in resource management regimes and, therefore, manipulation of nature will bring about greater control of production and increased output.

Sedentary marine resources such as mussels and oysters seem to offer excellent opportunities for the development of sustainable resource use under certain types of management regimes. Shellfish stocks can be assigned to specific owners and user groups and cultivation or semi-cultivation is possible by collecting oyster brood or mussel seed and replanting these on plots that provide optimum ecological conditions for growth and reproduction. In theory, the owners-cum-culturists will reap the benefits of good governance. As James Acheson hypothesises: 'where property rights exist, there would be less likelihood of overexploitation of resources, larger catches, more efficient use of capital, and higher wages to fishermen' (1981: 301). Therefore, the incentives to invest in governance structures are strongest among owners and weakest among authorised users (Schlager and Ostrom, 1992: 257).[1]

The idea, and the practice, has a long history that goes back centuries. In the Netherlands, it was from the mid-nineteenth century onward that the cultivation of European flat oysters (*Ostrea edulis*) and blue mussels (*Mytilus edulis*) on privatised plots was attempted at a large scale. This comparative case history attempts to throw light on developments in the oyster and mussel fisheries in two distinct settings in the Netherlands. It devotes special attention to the transition from capture to culture fisheries, which failed in one place (the isle of Texel), but succeeded in another (the province of Zeeland).

Detailed case histories covering a fairly long time span can throw light on the ways in which people understand their natural and social environments and how they relate and adapt to them. They can also show the diversity and complexity of adaptive dynamics in maritime settings. Through comparison of these cases, it will become clear that there are no easy solutions to complex problems. It should make those who believe privatisation to be an easy solution aware of the intricacies of the enclosure of the marine commons.

Complexity, Diversity and Dynamics: A Conceptual Appraisal

Fisheries are economically, socially and culturally complex, diverse and dynamic systems of interactions between humans and the natural environment (cf. Hamilton *et al.*, 1997). Often enough, however, fisheries management deals with single fisheries, reducing the complexity factor, while ignoring the factors of diversity and dynamics. These factors are often regarded as 'nasty' complications. Dealing with socio-political situations and developments 'as if' they were simple, homogeneous and static indeed provides for easier management tools (Kooiman, this volume). If you forget that you have simplified the issue, serious complications may be the upshot of resource management schemes.

Therefore, it seems appropriate to dwell a little longer on the concepts of complexity, diversity and dynamics in order to fully understand and appreciate their scope and importance. Fishing is an 'evolving system', a historical, economic and political process (Durrenberger and Pálsson, 1985: 120). Therefore, we need to take a diachronic perspective, explicitly in some instances, devoting attention to endogenous and exogenous forces impinging on the system or subsystem. There can be intricate patterns of relationships between forms of resource exploitation and the socio-cultural composition of communities, making for quite diverse ways of humans interacting with the biophysical environment. In that connection, the homogenising view of people's behaviour inherent in 'Tragedy of the Commons' scenarios grossly underestimates the importance of socio-cultural diversity. The use of communal natural resources in complex, diverse and dynamic socio-ecological systems cannot simply be explained by such simplistic and deterministic models as the 'Tragedy of the Commons' model. It should be interpreted in a much broader contextual framework. Though this will certainly complicate things for the student of common-pool resource utilisation, it would be unwise to simplify for the sake of parsimony. Besides being an oversimplification, the social consequences of departing from such a model are enormous and probably irreversible. It is well known that human behaviour, including conscious behavioural strategies, often have far-reaching unforeseen and unintended consequences. The same goes for fisheries management. Therefore, it is pertinent that we devote attention to the wider context of the fisheries and make sure that we incorporate as many contextual factors as possible in the models underlying fisheries governance structures. However, it is still important to allow for flexibility, lest rigidity hinder short-term responses to management failures. Enabling adaptive performance is a key issue here.

Adaptive strategies and processes result from positive and negative feedback loops. Adaptive strategies involve conscious decision-making. Adaptive processes are feedback loops operating outside of cognitive awareness. Adaptive dynamics are the sum total of strategies *and* processes (cf. Bennett, 1976). There are individual and collective adaptive strategies. Sometimes, these crosscut each other, giving rise to tensions that in turn may develop into conflicts. As the heterogeneity of a group of resource users increases, and as resource constraints increase, 'use rules' may become more difficult to maintain (cf. Runge, 1986: 630). An irreversible transformation of the system of resource utilisation - for example, through enclosure of the commons - may occur. It is this type of transformation that this Chapter will focus on.

Claiming the Commons: Texel Oysterers

To the southeast of the Dutch Frisian Isles, in the western part of the Wadden Sea and the northern part of the Zuyder Sea, oysters and mussels were caught from at least the early 1700s onward. The fishermen of the isle of Texel caught or gathered considerable amounts of oysters in the waters near their island, using dredge nets or small rakes when the receding tides left the flats exposed.[2] In principle, entry to these waters was free to all and so was the exploitation of resources in them. By the mid-eighteenth century some 60 Texel vessels and another 85 from the islands of Schiermonnikoog and Terschelling were involved in oyster fishing. Although access was open, Texelians claimed special use rights to certain locations in the vicinity of their island, which they considered communal grounds. The islanders replanted the oysters they had gathered or caught in the public domain on these plots, which were located in a shallow cove on the island's northeast side. Each oysterer had a parcel, demarcated with branches on the corners. The bivalves were fished and gathered in the public domain (*res publica*) throughout the year. These mature and immature oysters were then replanted on the plots, where they were tended and cared for until they could be marketed.

Thus, part of the waters of the western Wadden Sea and the northern Zuyder Sea was considered *res communis* and there was a quasi-form of oyster cultivation. Even when the fishery was free in a formal sense, there were informal regulations in which access and usufruct of these plots by Texelians were arranged, and agreements to exclude outsiders: each plot was 'habitually respected as someone's property'[3]. The Texelians' ambition to claim access to and use of certain locations for themselves amounted to the exclusion of outsiders; in other words, to obtain 'privileged space' (Acheson, 1981: 281), not so much to protect or conserve the resource. However, the communal use and management of the nursery beds had the unintended consequence of advancing oyster reproduction. The cove where these beds were located provided excellent conditions for oyster reproduction and growth. Moreover, the technological means available to fishermen were rather simple; gear efficiency was concomitantly low; often oysterers could not sail due to storms and ice drift, they did not market oysters between April and October, and they refrained from sailing on Sundays. Moreover, the vessels had but a small range of action.

By and large, this complex system of capture and culture fisheries worked out well. There appeared to be no excessive fishing until the mid-1840s. The Texelians shipped from one to eight million oysters annually. However, from then on, catches declined year after year and the oyster banks seemed nearly exhausted. To keep their trade going, Texel oysterers began to import oysters from France and England and they also had to fish oysters in Zeeland waters to stock their plots. Catches in the vicinity of the island declined from several million oysters to a few hundred thousand. The image of an inexhaustible supply of oysters faded rapidly when the oyster crisis persisted in the following years. Eventually, this crisis even turned out to be an irreversible tragedy.

What caused this tragedy? A number of factors seem to have been at play, one reinforcing the other. In the 1820s and 1830s, the oyster banks were very productive. Oysters could be distributed to markets further afield with the rise of steam navigation and railroads. But increased catches at the same time meant lower prices and falling incomes. To keep their earnings at an acceptable level, the Texelians were forced to harvest even more oysters. The number of vessels and fishermen rose quite sharply between 1836 and 1846: from 60 to 80 boats, each manned by a crew of three. In the same period, the catching technology also changed. The dredge nets were improved and used more widely and more often, replacing the small rakes. Although total catches increased initially, catches per boat declined - a fair indication of excessive fishing.

State officials began to worry and considered measures to protect the oyster banks. However, the state did not intervene. It was afraid that regulating one fishery would disadvantage another. In the 1850s the state had appointed a committee to investigate the state of the fisheries, and this committee proposed to leave all fisheries unregulated. The Texel oysterers themselves were opposed to any regulation because oyster prices, which had risen again due to the scarcity, kept their income up to the desired level. It was precisely because of these high oyster prices that the fishermen marketed all oysters they were able to harvest, both mature and immature ones. But higher prices could not make up for lower catches, and eventually the income of Texel oystermen fell. Although their behaviour was damaging to fishermen as a collective, it was perfectly rational for each individual to catch as many oysters as he could. The mechanism of subtractibility applies: almost all of the gain would go to each individual fisherman, whereas the costs (over-fishing and ultimately exhaustion of the oyster banks) were passed on to the collective of users. The fishermen were also caught in a zero-sum game: if a fisherman would throw immature oysters back into the sea, another might catch and market them.

But the decline of the oyster fishery cannot be attributed to the fishermen's behaviour alone. Natural circumstances also contributed to it. Oysters are very sensitive to changes in the ecosystem. Even slight fluctuations in water temperature, salinity, seabed features and food supply can cause considerable mortality. Severe winters caused marked oyster mortality, and cold summers had a negative impact on reproduction. Moreover, storms and changing currents also had consequences for the oyster staple. More importantly, a land reclamation project in 1835 resulted in the loss of more than three-quarters of the cove situated on Texel's northeast side as a location for replanting and tending oysters. Texel fishermen could henceforth only deposit their oysters in what was left of the cove which, to make things worse, silted up. Thus, the natural *milieu* for the reproduction of oysters deteriorated.

This ecological deterioration is important, since the increasing scarcity implied that, relatively speaking, the level of exploitation of the oyster staple rose, for initially catching efforts did not decrease. To eke out a living, the fishermen had to catch as great a share as possible from the declining stocks. This mode of behaviour had little to do with an innate rapacious mentality, but everything with the fact that the fishermen's economic existence was endangered. Debts to shopkeepers and suppliers had to be paid and the costs of living met. The fishermen's short-term interests indeed prevailed, not because they were purblind and greedy per se, but because other options were lacking as yet.

But the oyster fishermen did not seek to continue their activities until they had caught the very last oyster. They were diligently looking for alternatives. They could no longer exist from oyster fishing alone, and the Texel fishermen shifted away from the pursuit of oysters to other fisheries. They still caught oysters but only during a short season, and then only as a marginal part of a varied seasonal cycle in which they switched between various fisheries. Thus, they opted for diversification. From the 1850s onward, fewer and fewer fishermen pursued oysters.

Some efforts were made to counter the oyster crisis. Following the example of French oysterers in Arcachon, attempts were begun to farm oysters on leased plots near the isles of Texel and Wieringen. Since 1859, this was first done on the initiative of the Board of Sea Fisheries (*Collegie der Zeevisscherijen*), and later by three private individuals hailing from Amsterdam. This form of oyster farming differed from the Texelian system of quasi-cultivation in that the lessees tried to catch spat using 'collectors' (usually shells) to which the spat could attach and grow. Previously the Texelians had only gathered or fished young bivalves and replanted them on parcels they had staked out. However, the experiments with oyster farming failed for several reasons. Severe winters, storms and deteriorated ecological conditions caused poor results, and on top of that oysters were frequently stolen from the plots because of insufficient policing. Most Texel fishermen did not regard this form of privatisation as an attractive alternative. Despite their often destitute situation, they were still opposed to government intervention. When the Board of Sea Fisheries in 1859 asked the Texel fishermen to voice their opinion on oyster farming, they answered: 'The Texel oysterers are very satisfied with the destiny which is afforded them by nature, and they also think that no one is powerful enough to lay down the law for nature in this respect.'[4] In their Calvinist world view, they perceived nature as a God-given entity in which earth dwellers should not intervene except through hard labour. Only a small number of Texel oyster traders perceived advantages in oyster farming, but they have never been very successful.

From Capture to Culture: Oystering in Zeeland

In the southern Dutch province of Zeeland, fishermen applied similar strategies to claim exclusive use rights to certain fishing grounds as Texel fishermen did. But here, territoriality led to conflicts and clashes between fishermen who either claimed the same locations for their exclusive use or did not respect others' claims and encroached on what others believed to be their 'communal grounds'. As a report stated about the situation in the early nineteenth century: 'it was the denial of this usufruct which caused most quarrels'[5]. Due to developments that were to some extent similar to the ones described for the Texel case, oyster catches diminished. The decline continued even after the State expropriated the Zeeland estuaries in the 1820s and assigned the fishery management to the Board of Fisheries for the Zeeland Streams (*Bestuur der Visserijen op de Zeeuwse Stromen*). This Board regulated fishing-gear, methods, seasons, and the size of marketable oysters. Yet these measures did not prevent or stop overexploitation. By the mid-nineteenth century similar problems beset oyster fishermen the world over: 'the natural banks were close to exhausted ... in most ... places where there had been any commercial exploitation of the oyster' (Clark, 1964: 43). Oysters are particularly prone to overexploitation because they are an immobile species and thrive in shallow waters where they can be harvested fairly easily.

However, in Zeeland waters ecological conditions (water temperature, salinity, the availability of food - i.e. phytoplankton) favoured the reproduction of oysters. The shallow and relatively warm Eastern Scheldt basin was an especially productive location. Though experimentation with oyster farming near Texel had not been very successful, some wealthy urban capitalists were undaunted and tried to lease plots in the Eastern Scheldt basin from the state. They had studied oyster cultivation methods in the French Bay of Arcachon and intended to apply similar methods in the Eastern Scheldt basin. In 1870, they succeeded in convincing the state to enclose a large part of this basin, which was divided into small plots. These plots were leased to the highest bidders at a public auction for ten-year periods with the possibility of renewal for another five years.[6]

This state measure brought an end to time-honoured and deeply rooted systems of customary tenure. Oyster farming required more security of underwater tenure and the state perceived the lease system as its 'rational' economic interest, granting ample opportunity to the forces of capital to capture the commons. In an agrarian society like the Netherlands, the idea that the productivity of tenure-based farming would by far exceed that of common-pool resource exploitation easily gained acceptance in state institutions and among its representatives. For example, the State fishery advisor P.P.C. Hoek was an early advocate of the enclosure of the marine commons: 'It goes without saying that an owner (even a temporary owner) will care more about the maintenance or even growth of the value of his oyster grounds than can be expected under a system of common exploitation. Under the latter, each fisherman will strive to catch as many oysters in as short a time as possible, being convinced that each oyster he will leave in the sea will not lead to his future prosperity, but will only extend the profit of the one who will come next and catch it' (Hoek 1878,: 390-391). However, it was not the fishermen who reaped the benefits of oyster cultivation, but mainly newcomers investing in the industry.

The independent Zeeland oysterers vehemently contested the enclosure of the commons – but to no avail. Many wealthy newcomers, most of whom were not fishermen but urban capitalists looking for investment opportunities, bid with alacrity and succeeded in obtaining the majority of the plots. Consequently, most oysterers saw themselves excluded from the best locations and had to find employment with one of the newly established oyster companies. Though they lacked the capital to work independently, they possessed the sailing and dredging skills the newcomers needed. Those who cherished their independence exploited the still free grounds or turned to musseling, a far less capital and labour-intensive enterprise than oyster farming. Obviously, the political process of defining and enforcing property rights was socially divisive because of its distributive implications (cf. Libecap, 1989: 4). Capitalist entrepreneurs from without became the captors of the locations that until the introduction of the lease system had been commons. As a consequence incipient class divisions came to show, a consequence which also followed the enclosure of the commons elsewhere (e.g. Taylor, 1983). However, there are few instances of local industries that were captured so immediately and totally by outside entrepreneurs as the Zeeland oyster industry.

Following the privatisation of oyster fishing grounds, the Zeeland town of Yerseke became the centre of oystering. Most entrepreneurs established their companies here because the town was near the privatised plots and linked to an international railway network. The technique of oyster farming was laborious; the entrepreneurs used limed roofing tiles ('collectors') to obtain spat. These tiles were placed in inter-tidal zones in the summer, the spat which stuck to the tiles had to be removed, and before the oysters reached a marketable size after six years, they had changed many hands and had been dredged up and replanted several times. Oyster farming proved a success. Before 1870, the number of marketed oysters hardly ever exceeded one million. By 1875 it was approximately 35 million. Still, supply could not keep up with demand, prices remained high and investors in the industry made considerable profits. Many were attracted to the oyster industry and at ensuing public auctions of plots the lease fees skyrocketed because prospective lessees began outbidding each other to gain access. By 1886 nothing remained of the free oyster fishery.

Among the newcomers were some of the Texel oyster traders who had previously attempted to farm oysters near the island. Though their attempts there did not yield good results, they were convinced of the advantages of the privatised system. Therefore, they moved to Zeeland where ecological conditions were better. Near Texel itself there were new efforts to stimulate oyster farming following the Zeeland successes. In 1884, the state also introduced the lease by public bidding in waters to the island's south. Nearly all the lessees were successful and wealthy Zeeland oyster farmers.

The lease system contributed tremendously to the boom in production and to the industry's capitalisation. At the same time, it led to a transformation of the social relations of production. From a relatively egalitarian business - all oyster fishermen were independent, had equal access rights and possessed similar means of production - it became strongly stratified. In the process, many formerly independent oysterers were proletarianised. Capital replaced labour as the most important factor of production.

Yet it was the transformation in the mode of production that led to problems. The large planters, companies and shippers as well as the newcomers of the 1880s had unreasonable expectations. In their competitive struggle for plots, they lost sight of potential risks. They restricted their view to earlier experiences and were too optimistic about the future. Hence, many over-invested, especially companies which were financed by extra-local shareholders who hoped to make quick money. Due to the heavy lease burdens, considerable labour costs and increased bivalve production, the high profit margins began to shrink or even turned into losses. With a meanwhile saturated market, the industry was assailed by a prolonged depression. The crisis was exacerbated by a deterioration of the oysters' quality caused by overproduction and severe winters. The oyster stocks exceeded those that could be sustained by the amount of phytoplankton in the Zeeland estuaries. A fishery biologist exclaimed: 'Oyster farming is cultivation, not fabrication'. The entrepreneurs tried to save on costs by switching from tiles to cockle shells as 'collectors' - a far less labour-intensive method. It was too little, too late. Many companies and large planter-shippers went bankrupt or withdrew from the oyster industry and a dwindling number of newcomers tried their luck in it. By the turn of the century, the image of oyster farming as a lucrative occupation had vanished.

Near Texel, oyster farming never really took off. By 1886 - two years after the introduction of the lease by public auction - most Zeeland lessees had given up their attempts to farm oysters near the island. The ecological conditions were too poor, the leaseholders did not give sufficient care to the plots, and policing and supervision were inadequate. The severe winter of 1890-91, which had also caused serious problems in Zeeland, dealt a lethal blow to oyster cultivation near Texel. Five years later, oyster farming was abandoned there altogether and under mounting pressure from Texel and other fishermen the privatised plots were returned to the public domain. A dwindling number of fishermen continued to dredge oysters, until after the building of an enclosure dike in the Zuyder Sea in the early 1930s the bivalves disappeared completely. In fact, most Texelians had switched to other fisheries long before.

In Zeeland, there were participants who benefited from the withdrawal of outside investors in the oyster industry. With fewer people interested in farming oysters, the lease fees dropped. This enabled petty planters and family firms to obtain a greater share of the plots. Family labour provided a 'shock absorbing capacity'. By curtailing consumption and/or expanding production, these planters succeeded in surviving bad times. For them, the rationale of capitalist production for the market did not imply that they quit as soon as their firms suffered losses; they would try to weather a depression as long as they could eke out a subsistence. In their worldview, oystering was as much a way of life as a way of making a living.

Domestication: Mussel Fishing into Mussel Farming

The versatility of family firms was also apparent in the Zeeland mussel industry[7]. In 1865, the Zeeland Board of Fisheries privatised several mussel banks in the Eastern Scheldt and other Zeeland waters. The Board demarcated plots and allocated these for the duration of ten years to cocklers by the drawing of lots. It also provided for police patrols to prevent theft and poaching. Henceforth, mussel fishermen gained exclusive access rights in return for a modest rent of a few florins. The plots were re-allotted on a ten-yearly basis. Capture fisheries gradually turned into culture fisheries, though there were still grounds where free mussel fishing was permitted. However, by 1886 all Zeeland locations suitable for mussel farming had been privatised. Only the fishing of mussel seed that was replanted on rented plots remained free. Besides, the cocklers could still catch mussels in the Zuyder and Wadden Seas, where access was open, and replant these mussels on Zeeland plots. The transition from fishery to semi-culture led to an increase in output, but did not bring about dramatic changes in the social structure of the occupational community of cocklers and labour remained the most important factor of production. The economic risks were smaller, but musseling was not as lucrative as oystering. Therefore, the big capitalists refrained from investing in the mussel industry.

A number of Texel fishermen specialised in mussel fishing or temporarily caught mussels as a part of their annual fisheries cycle, switching to other species when these fetched comparatively better prices. Unlike the Zeeland cocklers, Texelians did not sell their mussels in the consumer market. Instead, they exported mussels to England, where they were used as bait in the offshore fisheries. Texel mussel fishing took flight between 1873 and 1890, when the Texelians landed between 3.5 and 13 million kilograms of mussels annually. Though some Texelians attempted to establish market outlets in Belgium, they were unsuccessful since Zeelanders already operated there. After 1890, the demand for 'bait mussels' diminished sharply and most Texelians gave up musseling for the time being.

Following the growth of the Zeeland oyster industry, the number of cocklers also increased. Given the lower capital investment required, many former oyster fishermen, labourers and newcomers turned to musseling. Whereas the oyster industry became strongly stratified, the occupational community of cocklers remained fairly egalitarian. All mussel fishermen operated independently in family firms, possessed similar means of production, and had equal opportunities to rent plots by participation in the drawing of lots. Even though the profits were considerably smaller than those to be obtained in oystering, those who possessed little money but valued their independence became cocklers. Since the vessels were relatively small and cheap, it was feasible for every crew member, given reasonable luck, arduous labour, and a degree of thrift, to aspire for his own boat. Many mussel farmers were also fishmongers; they sold their bivalves mostly in Belgian cities. Several planters combined musseling and oystering when it became financially feasible for them to rent oyster plots.

In the early decades of the twentieth century, fierce competition for a share of the market resulted in ongoing overproduction of mussels. A similar process had also occurred in the oyster trade. Given the imbalance between supply and demand, prices dropped. As a result, most cocklers tried to increase production to maintain or improve their standard of living. This solution to the 'peasant dilemma' of course only exacerbated their situation. A brief recovery followed during the First World War when the demand for mussels seemed unlimited. Even Texelians took up musseling again during this period. But after the war ended overproduction was the rule once more. Things became even worse when due to the motorisation of the fleet the supply of mussel seed shipped home from the Wadden Sea increased. Many cocklers were quick to adopt the new technology of mechanical power.

Interlude: Comparative Analysis of the Capture to Culture Transition

In comparing the Texel and the Zeeland cases, it is striking that Texel oysterers developed a kind of quasi-cultivation of oysters in the early eighteenth century, whereas Zeelanders continued to catch the bivalves in the watery commons. Nonetheless, the oyster stocks in Texel's vicinity were nearly exhausted by the 1840s and were never to recover from this blow. Several factors contributed to this decline. Ecological changes were important, as were the effects of human agency in the form of greater gear efficiency, a growing fishing fleet, and market expansion, leading to an increased extraction of the resource. The situation was exacerbated in subsequent years when sharp price fluctuations and impoverishment and indebtedness of the Texel fishermen led to intensified harvesting.

The introduction of oyster farming in Zeeland as from 1870 was - at least initially - quite successful, but attempts to introduce oyster farming near Texel were hardly successful or failed altogether. What seems to have caused this failure in the Texel case is an irreversible deterioration of the ecological conditions for oyster reproduction. Even large-scale attempts to replant immature Zeeland oysters on Zuyder Sea plots that had been very productive until the 1840s, failed. Though I do not mean to reason destructive human behaviour away, I think changing currents and water temperatures and silting up of some locations were important factors as well. Even when the pressure on the resource was alleviated because most Texel fishermen diversified their activities, switched to other fisheries or abandoned fishing altogether, there was hardly any improvement in the oyster stocks[8]. I believe this to be a fair indication of structural changes in the natural environment.

Ecological conditions were much better in the Zeeland area. Here, oyster farming was quite successful in the initial stage after the transition from capture to culture fisheries. That is, human intervention in nature brought about a larger production of oysters. It would seem that there was an increased control of nature. But success had its paradox. Soon, production went beyond the carrying capacity of the Zeeland inlets, and the quality of oysters deteriorated. Nature had set its limits to cultivation. On top of that, over-investments and market gluts led to problems for many oyster farmers who had been newcomers to the industry in the 1870s or 1880s. A large number gave up oyster farming again in the 1890s.

The enclosure of the commons had formidable social consequences, too. The marine commoners were excluded from the marine domain. The lessees had to pay dearly for entry and use rights, and fishermen usually lacked the capital to become leaseholders. They lost access and became wage labourers for the newly established companies or for the large planter-shippers. Consequently, the enclosure of the commons led to the marginalisation of the original fishermen and to the rise of a maritime proletariat. Women and children were also recruited as wage workers. Unlike their Texel compatriots, Zeeland oysterers had fewer opportunities to diversify their activities. Almost all of the Zeeland waters were enclosed by the mid-1880s, so alternatives were few. As we shall see shortly, even musseling became a privatised business. In contradistinction, access to a large proportion of waters near Texel remained free and the island fishermen had ample opportunities to switch target species. It was only due to the oyster crisis and the concomitantly falling lease fees that former oysterers or their offspring were able to regain the entry and use rights of oyster plots. They usually worked with agnate kinsmen aboard ship and household budgets were often augmented by the income of female family members. This 'logic of the peasant fishermen' enabled them to withstand fluctuations in the industry and to continue oyster farming.

Compared to oyster culture, musseling was far less labour and capital-intensive. The required means of production consisted of a boat and relatively inexpensive gear. The fees for the rent of mussel plots remained modest. In contradistinction to the oyster trade, the mussel industry did not undergo a phase of rapid capitalisation because the monetary rewards were smaller and plots were not up for public bidding but allocated by lot. Besides, a free mussel fishery was permitted in the Zuyder Sea and Wadden Sea. Nonetheless, overproduction also began to hamper the mussel industry in the early decades of the twentieth century.

Self-Regulation and State Intervention in Shellfish Farming

Although the state had privatised large parts of the Dutch, and especially the Zeeland inshore waters, to enable oyster and mussel farming, it refrained from intervening in the industries concerning production. In this respect, the state believed in and pursued a *laissez-faire* policy. It was up to the planters and companies themselves to arrive at some form of regulation or to let the market's 'invisible hand' rule. In the 1910s and 1920s, both oyster and mussel farmers experienced a need to limit their output and through voluntary associations they attempted to arrive at a collective agreement to do so. But time and again, these agreements were undermined by free riders who did not join organisations established with this aim and by those who did join, but evaded the organisations' rules and regulations. It became clear to most planters that self-regulation would only work if an external authority enforced and supervised the rules. This happened in the 1930s, when mussel and oyster planters and shippers experienced the consequences of the economic crisis in the capitalistic world. On top of that, Zeeland oyster shippers were ousted from the English market by French and English oysterers. In response, the state established the Dutch Fishery Marketing Board (*Visscherijcentrale*), and the planters and shippers had to join this state organisation. Among many other measures, it set quality standards, quotas and regulated prices for mussels and oysters.

In the oyster industry, there was an additional reason for the State to intervene. A serious outbreak of shell disease and the proliferation of the slipper limpet (*Crepidula fornicata*), a food competitor, caused huge problems and decimated the oyster population. Both plagues were facilitated by the presence of huge quantities of cockle shells, which were scattered on the plots by the oysterers to collect spat. These shells hardly disintegrated and were a seat of diseases. The government imposed a ban on the use of cockle shells and imported French yearling oysters to aid planters who had lost a large percentage of their stocks. In order to secure a sufficient supply of oyster spat and yearling oysters, the state provided financial support to unemployed labourers who wished to collect brood by using limed tiles. The local government supplied them with roofing tiles and credit, while the state gave them access to plots near the shore which had not been used since tile farming was abandoned in the late nineteenth century. The reintroduction of tiles as collectors was hardly successful. The leading oyster planters began to use mussel shells as collectors and they also imported young oysters from Brittany to stock their plots. This considerably reduced the bargaining power of the petty 'tile farmers'. They usually had to accept the low prices offered by planters and planter-shippers that shoved the risks of the trade on to the petty oysterers. As one man stated: 'We could only lease those plots that were suitable for catching spat, not those fit for growing marketable oysters. So, eventually, you just had to sell. And there were many of us. When supply was large, prices were low, you just had to accept that. If you didn't want to sell for the money a planter offered, he would say: "Well, you'll come back when you're hungry"'. Meanwhile, the situation in the oyster industry had generally turned for the better and the state soon repealed its price regulations.

This was not the case in the mussel industry. The Fishery Marketing Board set minimum prices for mussels for export. The home market remained free, however. Soon Belgian dealers began to work with Dutch middlemen to evade the price regulations. To counter this situation, in 1935 the Central Sales Bureau of Mussels (*Centraal Verkoopkantoor van Mosselen*) was established, partly at the insistence of the planters, who suffered most from the evasion of the price regulations. Henceforth, all transactions between planters and shippers had to be made via the Bureau. Subsequently, it set quality standards and introduced fixed prices, both for mussels the Bureau bought from the producers and for the bivalves it sold to the dealers in turn. Moreover, it regulated the admittance of newcomers in order to curb the expansion of the number of mussel culturists and introduced a licensing system for shippers, thus reducing the number of cocklers who were allowed to ship their own merchandise. This management regime was still not quite successful; soon a new boom in output followed. In 1938, the Bureau responded by allocating production quotas, known as standard capacity numbers (*standaardcapaciteitscijfers*), to all individual cocklers, based on their estimated production in earlier years. Alternately, each planter was allowed to supply a certain quota to the Bureau. This rigid regulation of the industry, aimed at balancing supply and demand, proved adequate and the position of the planters improved. It had a stabilising influence, though it also brought about a fixation of the industry's structure and limited the expansion of individual firms. The quotas were fixed and non-negotiable. The only way to expand a firm was to buy another firm. The number of cocklers who kept sailing to Belgium started to fall, not only due to restrictions imposed by the

Bureau, but also because the transportation of bivalves was gradually taken over by trucking companies.

The oyster and mussel firms that had survived the economic crisis and other problems of the 1930s were faced with the consequences of war and occupation in the first half of the next decade. Many boats were confiscated, damaged or destroyed, fuel soon became scarce, musseling and oystering came to a near standstill, and export was impossible. The German occupying forces demanded the best part of the landings. In the oyster industry, the Germans replaced the lease system by a fixed yearly rent, calculated in terms of the estimated value of the plots. It was further regulated that the entry rights could not be transferred to other oysterers, as was the practice heretofore, other than by the agreement of the Secretary General of the Department of Agriculture and Fisheries. Lastly, the allotment of plots became based on the need of individual oysterers and companies. The Dutch government adopted these regulations after the liberation of the country in 1945. It also reduced the rent of oyster and mussel plots to stimulate the shellfish industry's recovery and to enhance the food supply in the Netherlands. Gradually, the industry managed to recover from the disruption of these years, although the position of small planters continued to be difficult. The organisations of planters and shippers gained a foothold in state-level fishery institutions so that they could defend their interests. Potential newcomers to the oyster industry could only gain entry if a firm relinquished its plots, a measure that had already been in place in the mussel industry since the late 1930s. The mussel trade continued to be much more tightly regulated than the oyster industry.

Myticola Intestinalis: A Blessing in Disguise

Following two good years, things appeared to get even worse. In 1950, a parasitic copepod, *Myticola intestinalis*, killed a large proportion of Zeeland mussels. Some cocklers lost over eighty per cent of their stock. The shippers were consequently unable to supply customers. The planters and dealers were powerless against this ecological disaster and feared that it presaged the end of musseling in Zeeland. Paradoxically, however, this catastrophe preluded a phase of capitalisation and expansion. Some enterprising planters gained permission to cultivate plots in the Wadden Sea, until then a location mainly used for seed fishing.[9] Soon all Zeeland cocklers relocated parts of their production areas to the Wadden Sea. Moreover, the mussel parasite vanished from the Zeeland inlets within a few years. Thus, there was an enormous expansion of the total area of plots available, which gradually increased from 4,000 to 10,000 hectares. Since the demand for mussels had also risen, the

Bureau considerably extended the individual quotas.

But other problems loomed large on the horizon. In 1953, a flood disaster struck Zeeland that was to have grave consequences for the oyster industry. Five years later, the government decided to dam up all inlets but one in the province. This would render mussel and oyster cultivation impossible. The Eastern Scheldt was scheduled to be shut off from the North Sea in the 1970s. In anticipation of the Zeeland delta being dammed up, the relocation of mussel farming to the Wadden Sea was speeded up.

However, growing opposition by fisher folk and environmentalists led to a reconsideration of the government decision to dam up the Eastern Scheldt. In 1976, Parliament approved the construction of a storm-surge barrier that would maintain the tidal regime. This meant that mussel and oyster farming in the Eastern Scheldt would remain possible. Thus, the total available area for mussel cultivation increased, though the acreage of mussel beds in Zeeland decreased.

In the meanwhile, in 1967 some of the most successful planters and dealers had persuaded the Ministry of Agriculture and Fisheries to withdraw most of the protective measures that had been introduced in the 1930s. The quota system was abandoned and henceforth mussels were sold at a free auction in Yerseke. This did not imply that the industry returned to a *laissez-faire* situation. The State retained formal jurisdiction over shellfish grounds, it still polices the waters, monitors the sanitary condition of shellfish farming areas and finances a department of the Netherlands Institute for Fisheries Investigation in Yerseke, which carries out biological research and provides the shellfish farmers with information and advice. However, the involvement of the industry's participants increased. The Industrial Fisheries Board (*Produktschap voor Vis en Visprodukten*), an organisation for the fishing industry as a whole, together with representatives of all branches of the mussel industry - planters, dealers and canneries, united in the Mussel Advisory Committee (*Mosseladviescommissie*) - now determined quality standards and maintain minimum prices. In order to make this work, a fund (*Mosselfonds*) was created. The planters deposited a small percentage of each sale with this fund. If their mussels did not meet with the quality standards, or could not be sold at the bottom price at the minimum, they are compensated by the fund. The mussels are bought by the fund, planted on plots and sold at a later date. Thus, this system is quite flexible. The Industrial Board and Advisory Committee also negotiate with the Ministry of Agriculture and Fisheries regarding the replacement of plots that had become unproductive, for example due to silting. In general, this co-management regime has been successful so far. Production has boomed, but supply could not keep up with demand and, concomitantly, prices have increased sharply.

However, there were some disadvantages as well. The expansion of mussel farming in the Wadden Sea was at the expense of shrimp fishermen in the north of the country, who saw their shrimping territory drastically reduced. Some fishermen of the isle of Texel, for example, tried to gain permission to rent plots in order to start mussel cultivation, too. Their efforts did not bear fruit. The Ministry of Agriculture and Fisheries refused to give them access to such plots because Zeeland mussel planters had to be compensated for a loss of mussel beds as a result of the damming up of several inlets in the Zeeland delta. Most shrimpers, who only received nominal indemnification for the diminution of fishing grounds, bitterly resent the fact that Zeelanders plant mussels in what they consider to be 'their' territory. In recent years, there has been a scarcity of seed mussels, and the cocklers have exploited many natural banks to satisfy their needs. Environmentalists have waged a battle against the cocklers because they catch the mussel seed birds also prey on. Apparently, the environmental movement grants birds more rights than cocklers.

The expansion of the mussel industry also implied demerits for certain categories of cocklers. The relocation of many production areas to the Wadden Sea necessitated larger boats. This changed the balance of forces of production from labour being more important to capital becoming more important. A period of rapid modernisation, increases in scale and mechanisation ensued. These changes worked to the advantage of the large mussel culturists and to the detriment of the petty planters, who were unable to keep pace with the process of growth because they lacked the funds to modernise. Many could no longer compete and especially those without successors had to sell their business to large-scale planters and dealers. The number of firms decreased from 143 in 1960 to 80 in 1985. This development, encouraged by the Ministry of Agriculture and Fisheries, aimed at fewer but more profitable enterprises. Today, the state follows a very restrictive policy with regard to the admittance of newcomers and the number of participants in the mussel industry has hardly changed. Only those inheriting a family business or experienced employees who want to set up their own enterprise can get a license, provided that the total number of firms does not increase. Thus, the expansion of the mussel industry as a whole brought about the demise of small enterprises. Nonetheless, the industry's social organisation is still predominantly based on family firms.

The Demise of the Oyster Industry

The historical trajectory of the oyster industry took quite a different route than that of the mussel trade. The 1958 decision to dam up the Zeeland inlets had much more drastic and especially negative consequences for oysterers than for cocklers. Unlike mussel cultivation, oyster farming was only possible in Zeeland waters. As we have seen, the dike that enclosed the Zuyder Sea in 1932 implied that this marine domain turned into a fresh water lake where oysters could not survive. On top of that, ecological conditions in the Wadden Sea deteriorated following the dike's construction. So much so that by the 1930s the oysters disappeared from the Wadden Sea. Attempts to farm oysters there during this decade failed time and again. Whereas the cocklers could relocate their production areas to the Wadden Sea, this alternative was not available to the oysterers. Those who combined oystering and musseling were in a favourable position because they could invest more in musseling or switch to this branch of trade altogether. The state developed a compensation programme for the oysterers. They acquiesced in their fate, going about their work as usual, in an effort to make the best of the situation. There were still many years to go before the Eastern Scheldt would actually be dammed off.

Then, in 1962-63, an extremely harsh winter decimated the oyster stocks.[10] Only an estimated five per cent survived. This dealt a lethal blow to the majority of oyster firms. The bivalve producers and dealers suffered great financial losses and on top of that the prospect was that the Eastern Scheldt would be dammed off in the near future. Most oysterers deemed it senseless to continue their occupation. The majority of planters and all small tile farmers decided to quit and accept state indemnification, amounting to approximately fifty per cent of the real damages. Some retired, while others set up new ventures. The wage labourers could usually find industrial employment, since it was a time of rapid industrialisation in the Netherlands. The planter-shippers also collected financial compensation. Only some of them continued to rent a few plots. Since the native bivalves were virtually wiped out, they imported large quantities of four-year old oysters, replanted them and marketed them one year later. Because supply was scarce and competition minimal, they could make a comfortable living. The enterprise, however, was not without its risks. The bivalves, mostly imported from France, did not adapt to the lower Zeeland water temperatures in winter and mortality rates were high.

For this reason, the majority of those still renting plots began to refrain from using them. They consequently had to relinquish the parcels because a law passed shortly before forbade lessees to let underwater grounds lie fallow. Only ten planter-shippers persevered, and they cheaply rented the relinquished plots in addition to the ones they already leased. As a consequence, they gained access to extensive underwater grounds and accounted for 99 per cent of the oyster production. Thus, as a result of the ecological disaster following the winter of 1962-63 and the Delta Plan, the social structure of the industry that arose in the 1930s, suddenly came to an end. In contrast to the ten affluent planter-shippers who intended to continue oystering as long as possible - that is, until the enclosure dike was to shut off the Eastern Scheldt from the North Sea, the small tile farmers and petty planters had neither the means nor the energy to stay in the industry.

But as we have seen, in 1973 the government altered its policy and decided that the Eastern Scheldt would not be dammed up completely. Instead, a storm-surge barrier was to be built. This left the remaining planter-shippers in an extremely strong bargaining position. Firstly, they rented nearly all the oyster plots. Secondly, they had established an association and as the sole organised representatives of the oyster industry, they advised the civil servants of state institutions. Of course, they highlighted their own interests. Therefore, they virtually controlled the industry. Though some former planter-shippers (or their sons) attempted to regain entry to the plots, they were unsuccessful. The monopolists continued to play the game of import and export, seeking a quick turnover. But they had grown careless. Against the advice of fishery biologists, they imported and replanted French oysters that turned out to be infected by a parasitic disease, later named *Bonamia ostreae*. In 1980, it was established that this disease had affected the oyster stocks in the Eastern Scheldt. Thereupon, the state banned the cultivation of the European flat oyster in the inlet.[11] It was only in another inlet, the Grevelingen, that the cultivation of *Ostrea edulis* could be continued. In the second half of the 1980s, an annual yield of approximately ten million oysters was harvested there. However, in 1989 it was established that *Bonamia ostreae* had spread to the Grevelingen and since that year harvests had declined to 0.5 million oysters. This miserable situation has continued until today. For decades, the economic importance of the mussel industry has exceeded that of the oyster trade by far.

Conclusion

This Chapter has shown that captors can be turned into culturists. However, privatisation does not necessarily provide shellfish farmers with incentives to maintain their harvests at an ecologically sustainable level. Some of the consequences of privatisation of the Zeeland oyster banks were overproduction, resource deterioration, overcapitalisation, marginalisation of the original fishermen, the creation of social divisions and maldistribution of incomes. On the other hand, privatisation in the Zeeland mussel fishing and farming industry provides an example of successful fisheries management and sustainable use. Under certain conditions, certain forms of privatisation may be viable resource management instruments, but not all forms of privatised management regimes are necessarily successful. There are no easy solutions leading to the sustainable use of marine common property resources. As we have seen, one serious problem is not depletion of resources, but overproduction.

For each individual shellfish farmer it is 'rational' to increase production, and this can bring about overproduction at a collective level. If overproduction occurs, lower prices will be fetched in the market, creating incentives for individuals to produce even more, which can undermine the ecological carrying capacity of the marine environment. Following such experiences, shellfish planters have often opted for collective action to solve resource management problems. But time and again, these strategies failed as a consequence of free riding and evasion of the rules and regulations. Therefore, the planters asked the state to wield its authority and enforce the rules and regulations. But again, their attitude was ambivalent; when they perceived advantages in state involvement, they were in favour of state intervention; when they felt their freedom to act was stifled too much, they would try to circumvent the rules or ask the state to withdraw certain measures. However, the co-management scheme which developed in the mussel trade and in which the interests of the state, the producers and the dealers are balanced and constantly negotiated seems to have been successful so far.

Though mariculture, and the farming of shellfish in particular, implies greater control of nature - at least in comparison with capture fisheries - and increased production, a caution is in order. Increased control at the same time enhances vulnerability, since people become more dependent upon the resource being controlled and the means by which it is controlled. If successful, farming a single marine species will be promoted either by the producers and/or by state agencies and this can eventually lead to a kind of mono-mariculture. Shellfish farming often leaves little room for multiple use, and a successful branch of shellfish farming usually expands at the cost of other fisheries and other forms of mariculture, taking up much space formerly used by many other fishermen. Moreover, privatisation often leads to marginalisation of the commoners and protects the interests of participants once they have gained access. It also creates tremendous barriers for potential newcomers to the industry. These conflicting interests and equity problems have to be solved or frictions will ensue. Apart from these social problems, monocultures can lead to ecological and economic problems.

There are many examples of diseases assailing specific branches of the shellfish industry, for example *Bonamia ostreae* (or MSX) in oystering and *Myticola intestinales* in musseling. These diseases spread rapidly and can hardly be fought. Although the commons may be divided up for human use, carelessness of one or a few users or sheer forces beyond man's control can be destructive for the common resource and hence for the collective of users. When in 1980 *Bonamia ostreae* was discovered in the Dutch oyster industry, a ban on the farming of *Ostrea edulis* was installed and the plots lay fallow in anticipation of the disappearance of the disease. This situation has lasted for many years now, much to the dismay of cocklers who were not allowed to use this extensive area and at the time were in need of and applied for more plots, to no avail however. Therefore, it would seem unwise to place all one's eggs in a single basket. It is probably preferable to maintain diversity so that if a disaster or a disease strikes one species, it does not bring about a disaster of the magnitude often seen in monocultures.

For the same reason, switching species should not be ruled out beforehand. If lessees have to pay dearly to get access to plots and if they do not have alternatives, resource deterioration may not prevent them from harvesting their stock. They are likely to take the oldest year-classes necessary for the reproduction of a species. They may even have to, for their short-run economic survival. It is therefore necessary that an external authority - for example the State or some co-management body - closely monitors the industry and has the power to intervene in it. As Greenpeace International campaigner Mike Hagler warns, one should not be too optimistic regarding the viewpoint that farming the seas 'necessarily brings with it sound husbandry'. He goes on to state that '(e)nthusiasts of farming the seas should reflect that upon land, what has often grown back after repeated attacks upon wilderness has not been rich diverse forests, not even a sustainable monoculture, but degraded woodland, scrub, poor grazing land and ultimately desert' (1995: 78). Though Hagler's view is a grim one, it is worth reflecting upon. In this respect, the tragic story of *Ostrea edulis* speaks volumes. The dilemma, then, is that shellfish cultivation can enhance production, but only at the cost of ecological diversity. How to steer a clear course between this Scylla and Charybdis remains one of the major challenges for the management of shellfish farming in the future.

As John Bennett observes, 'human systems are not unitary, but are dynamic and proliferational: when needs cannot be satisfied by one system, a subsystem is likely to form through the adaptive actions of individuals; or, the individual may switch his behaviour from one system or subsystem to another, seeking out more congenial alternatives' (Bennett, 1976: 25). A common strategy of fishermen is to switch target species once it has become unrewarding to catch a species which was pursued previously; it is a 'normal tendency of fishermen to switch away from declining stocks' (Townsend and Wilson, 1987: 323). This may bring about a dispersion of pressure on marine resources, a consequence that was neither intended nor foreseen. If not tied to a single resource (for example when one 'owns' this resource as a tenant or as a person entitled to a certain quota), fishermen are likely to take optimal advantage of the variety of marine ecosystems, choosing to utilise niches as they see fit.

However, one should not consider these adaptations as cybernetic processes automatically leading to homeostasis. Nor should one mistake them for evidence of control over nature or signs of ecological wisdom. We are still in need of a theoretical perspective that can account for 'the shaping and constraining forces of ecological adaptation, but sees them as operating through systems of cultural meanings and social relationships, that sees internal conflict and contradiction within social systems, as well as adaptation to material circumstances, as dynamic forces' (Keesing, 1981: 171-172). These adaptations themselves are often quite diverse. As Orvar Löfgren rightly observes, 'ecological variations combined with local demographic and economic factors and changing national government policies have generated a diversity of adaptations among the coastal populations' (1979: 85-86). Socio-cultural factors could be added. Unless we take these integral dynamics into account - that is, the interdependencies and interactions of factors and actors in a system of resource use and the processes and transformations they bring about - we will be unable to get a full grasp on the complexity, diversity and dynamics of renewable resource utilisation.

Notes

1   According to Schlager and Ostrom's conceptual scheme, 'owners' have the rights of access and withdrawal, management, exclusion and alienation; 'proprietors' have all of these except the right of alienation; 'claimants' have the rights of access and withdrawal and management; and 'authorized users' only have the right of entry and withdrawal (1992: 252).

2   For a more comprehensive account of this case history, see Van Ginkel 1996.

3   *Verslag van de Staat der Nederlandsche Zeevisscherijen* [Annual Report on the State of Dutch Sea Fisheries] (1860: 36). Henceforth: *Sea Fisheries Report.*

4   Appendix to a letter, dated 23 January 1859, from the *Collegie voor de Zeevisscherijen* [Board of Sea Fisheries] to the Mayor and Councillors of Texel regarding the cultivation of oysters (Texel archives, no. K-853).

5   *Sea Fisheries Report* 1861: 16.

6   The transition from capture to culture oyster fishing and later developments in oyster farming is described in more detail in Van Ginkel (1988, 1989).

7   On developments in the Zeeland mussel industry, see Van Ginkel (1990, 1991a).

8   On these and other adaptive dynamics, see Van Ginkel (1994b, 1995).

9   This led to upheaval among those who were initially excluded from access to the Wadden Sea. On the contest between those who could go and those who could not, and the prolonged conflicts which ensued in the Zeeland community of Yerseke, see Van Ginkel, 1991b. On the idiom and ideology of Yerseke shellfish planters, see Van Ginkel (1994a).

10   Oysters are quite vulnerable to changes in the natural environment. Much more so than mussels, which can for example withstand low water temperatures for long periods of time. The 1962-63 winter had 71 consecutive days with water temperatures below minus 1.5 degrees Celsius.

11   That is, the cultivation of *Ostrea edulis*. Another species – known as the Japanese oyster, introduced in the 1970s - has proliferated in Zeeland waters and is exploited by 30 oyster firms and permit holders. In recent years, catches have gone up from approximately 5 million to 12 million. This is partly a capture fishery, partly a culture fishery. For a more comprehensive description of developments in shellfish cultivation after the construction of the storm surge barrier, see Dijkema (1988).

4 Co-Governance in EU Fisheries

The Complexity and Diversity of Fishermen's Organisations in Denmark, Spain and the UK

DAVID SYMES AND JEREMY PHILLIPSON

Introduction

The problems that currently beset the fisheries of Europe - and elsewhere - are the product of a governing crisis. They reflect the failure of conventional management systems that rely upon centralised, 'command and control' decision-making and top-down modes of policy delivery. Such systems are essentially based on legal and administrative instruments. Each new challenge is met with a tightening of existing rules, the introduction of new legislation and the promise of stricter enforcement. As a result, fishermen are increasingly burdened by externally imposed regulations that may appear inappropriate or irrelevant. Estranged from the policy process through weak levels of integration of the industry's representative bodies in decision-making, fishermen sense that they have become the object rather than the subject of the policy process. Commitment to and compliance with the stream of regulations is reduced.

The search for solutions to the moral dilemmas of ineffective regulation has shifted away from the 'content' of management policy and the selection of the most appropriate regulatory mechanisms to the reform of the institutional frameworks within which policies are framed and implemented and, in particular, to the realignment of relationships between the regulators and the resource users. The concept of co-management has come to the fore in the last ten to fifteen years, both as an embodiment of co-operation and, more elusively, as a concrete expression of an appropriate organisational structure for fisheries management.

There is a need to develop an agreed, clear and unambiguous understanding of the co-management concept. It will require defining the concept with precision, identifying the conditions under which co-management may prosper and laying down guidelines on 'ideal structures' for co-management institutions. There are two alternative approaches to this set of tasks. The first involves a comparative analysis of existing systems. We can analyse examples of successful co-management schemes and ask: how do they do it? Once we have correctly identified the salient features, we should be able to implant them into our own systems of management. But we would caution against this form of transplant surgery. One of the lessons to be learned from comparative analyses (Jentoft and McCay, 1995; Sen and Raakjaer Nielsen, 1996; Symes *et. al.*, 1996) is the overriding importance of the prevailing political culture and the organisational structures of the fishing industry. The body politic will quickly reject the transplanted co-management system unless it is fully congruent with the broader system of governance and the particular conditions of the fishing industry. The alternative approach is to build up the concept of co-management from first principles and this is the task of the first section of this Chapter. After such theoretical considerations will follow an empirical analysis of existing and potential systems of co-management in Denmark, Spain and the UK.[1]

Theoretical Considerations

Defining the Options: Alternative Management Systems

According to Sagdhal (1992) 'the concept of co-management is widely used but poorly defined'. Weak definitions, such as that which posits that co-management involves 'some form of institutional arrangement between government and user groups to effectively manage a defined resource' (Jentoft, 1989) or 'an arrangement where responsibility for resource management is shared between government and user groups' (Sen and Raakjaer Nielsen, 1996), lead inevitably to the inclusion of a very wide range of user group participation within the concept. While such definitions do clearly identify the essential element of a shared responsibility between the state and the user groups, they fail to define the precise nature of the relationship. This may be fine for the purpose of academic debate, but it is decidedly unhelpful in terms of a more pragmatic approach to finding appropriate solutions to the current crisis in fisheries management. The question is where to draw the line between co-management and other forms of user group participation.

It may help if we start by trying to locate co-management more precisely within a simple continuum of management systems. Five basic models can be arranged along a continuum (Fig. 1) from extreme centralisation of policy making and management functions to the complete devolution of those functions to an autonomous, independent, non-governmental organisation.

Enlightened dictatorship occurs where the central state defines and administers all aspects of fisheries management (policy formulation, implementation and enforcement), without direct negotiation with professional groups or regional bodies, and where the state also exercises legal powers for the enforcement of the management strategy. However, to be 'enlightened' it must have access to high quality information and opinion concerning, not only the biological state of the stocks, but also the socio-economic implications of their management and what may constitute acceptable management strategies among the user groups.

The source of enlightenment might be an Advisory Council. Essentially this is a device for incorporating various interest groups, both within and without the fishing industry, into the policy framing process along with the government's own scientists, economic advisers etc. through consultation. An Advisory Council might thus be used as the intelligence-gathering organisation in support of the enlightened dictatorship. But, by definition, it is a non-executive body; its role is to advise and not take executive decisions. It does not amount to co-management in that there is no sharing of responsibility for management. The central authority may choose to accept, modify or ignore the advice offered.

Fig. 1: Continuum of alternative management systems

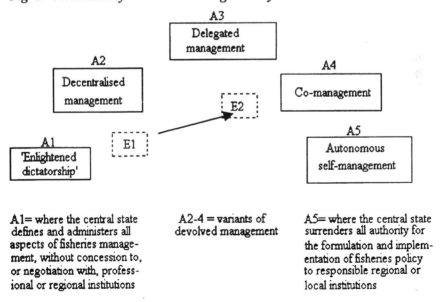

A1= where the central state defines and administers all aspects of fisheries manage-ment, without concession to, or negotiation with, profess-ional or regional institutions

A2-4 = variants of devolved management

A5= where the central state surrenders all authority for the formulation and implem-entation of fisheries policy to responsible regional or local institutions

E - E1 and E2 describe existing and possible future locations of the management system

Decentralised management involves the downward transfer of specified powers and responsibilities from the central to the local state. A limited form of decentralisation may simply involve the transfer of administrative responsibilities, but without powers of independent decision-making, to regional offices of the state bureaucracy or to nominally autonomous local authorities. Such responsibilities might include the administration of grant aid, inspection services and data collection.

A more significant form occurs where substantial powers of decision-making are granted to regional or local authorities, implying a transfer of responsibility for management of inshore waters on a regional basis, bringing management decisions closer to the appropriate user groups, but not necessarily implying their active involvement in decision-making. Decentralisation infers a sharing of management responsibilities between the central and local state. Discrete functions are reserved for the local state, but the framing of the broad policy initiatives will normally rest with the central state. Insofar as decentralisation involves the retention of management responsibility within the formal democratic system, it does not constitute an example of co-management.

Delegated management implies the surrender of powers and responsibilities for management to other organisations that lie outside the democratically accountable system of local, regional and national government. Delegated organisations take the form of agencies specifically designated by central government, either as QUANGOs or private, non-governmental organisations formed as trusts, co-operatives or private companies. A crucial feature of delegation is the system of public auditing or control of the independent authority. QUANGOs are nominally responsible to a government department; private organisations are more problematic. Their roles and responsibilities may be defined through legislation and the interests of the consumers or user groups placed in the hands of an independent regulator.

Delegation may occur at any level. An Executive Council, similar in structure to an Advisory Council but with executive functions, may be empowered to undertake national fisheries management. Similarly, regional or local management boards may discharge a range of management functions. As with decentralisation, there is likely to be a clear division of powers and responsibilities reserved separately for the state and the delegated agency. Thus the state will set limits within which the delegated agency has powers to act; such powers may be no more extensive than those ascribed to decentralised authorities.

Although delegation clearly satisfies the notions of devolved management responsibilities and user group participation, it does not necessarily involve user groups in the formulation of policy. Most delegated responsibilities refer specifically to the implementation of policy, previously determined at the centre, and may represent little more than a transfer of the burden of detailed administration from central government to an appropriate local organisation. The administration of sectoral quotas by producers' organisations in the UK may be seen in this light.

Co-management is, in effect, a special example of the delegation process where 'self-governance, but within a legal framework established by government, is a basic principle, and power is shared between user-groups and government' (Jentoft, 1994). It is distinct from other forms of delegated arrangement in the sense that it focuses on the direct involvement of user groups in an active management role. It thus goes beyond consultation between the state and representatives of the fishing industry in that the delegated institution, embodying user group interests, not only has a direct role in joint decision-making but also the authority to make and implement regulatory decisions in specified areas of responsibility.

As an example of co-partnership, it is important that the autonomy of both the centre and the delegated institution, in their respective areas of responsibility, is fully recognised and respected by both parties. Co-management is built upon the foundations of mutual respect and trust. Problems of unequal access to information, organisational resources and finance may threaten to disturb the balance of power. It is important, therefore, that the state should ensure that sufficient resources are available to the delegated institution to maintain its independence.

Autonomous self-management, or self-governance, represents the ultimate form of delegated management, a logical counterpoint in the continuum to enlightened dictatorship and the antithesis of management by the central state. However, it is doubtful whether self-governance is a realistic option for the management of modern fisheries, except perhaps at a very local level. Under self-governance, all responsibility for formulation, implementation and enforcement of fisheries management is devolved to a responsible organisation representing the interests of the user groups, without the intervention or collaboration of the state. The fishermen are in sole control of the fishery, organising its management through formal institutions designed and operated by fishermen's organisations alone. Self-governance, so defined, scarcely exists anywhere in contemporary Europe, where local systems of management have been largely overwhelmed by state bureaucracies. It may survive in historical fragments in Spain (*cofradía*) and Mediterranean France (*prud'homie*) but today these are enveloped within a state led institutional system.

More fundamentally, it is questionable whether such self-governance institutions would be competent to tackle the range of management responsibilities demanded by modern fisheries. Traditionally, they limited their management objectives to achieving distributional equity among a closed membership through systems of management closely akin to 'traffic regulation' schemes, deciding who should have access to which grounds, when and with what gears. They lacked the means - in terms of scientific knowledge of the state of the stocks, normally the 'property' of the state - to undertake resource conservation or stock management, which in any case was an inessential management objective while the level of fishing effort remained low. Modern fisheries management requires a partnership between the state and the user groups - each contributing a unique set of assets and skills - rather than the usurpation of responsibility by one party or the other.

Hopefully, this attempt to locate co-management within a broader spectrum of management options will have brought us somewhat closer to a more precise definition. Co-management involves two main functional ingredients: consultation between the central administration and the user groups over the content of the management strategy and the delegation of specific management functions to user group organisations. But neither of these is sufficient, on its own, to fulfil the essential conditions of co-management. This requires a more holistic approach and a more closely integrated relationship between the central administration and the fishing industry, in which there is a commitment to co-responsibility and co-operation throughout the policy process involving the formulation, implementation and monitoring of policy decisions. In other words, the relationship must be elevated from one that allocates a passive or reactive role to user groups to one where the user groups assume a more active or proactive role in management. This may require granting limited legislative powers to responsible and relevant user group organisations, through the provision of by-law capabilities. But co-responsibility is not meant to imply equal power sharing in policy matters. The balance of power will nearly always reside with the state, which is ultimately responsible for defining the scope and conditions of co-management. What is required is that the roles and responsibilities ascribed to user groups should be clear, specific, substantive and permanent. In particular, the contribution of the user groups should be seen as providing a constructive force, rather than a countervailing power.

Although the aims and objectives for the co-management of fisheries are no different to those set for any other form of management system - biological sustainability, economic efficiency and distributional equity - the manner of their achievement will be profoundly different. As a result, the benefits claimed for a co-management approach are quite extensive, including *inter alia*:

➢ a more open, transparent and less hierarchical policy system;
➢ a broader basis of information and knowledge, wherein scientific information can be complemented by practical experience;
➢ an increased rationality for the regulatory system, as co-participation helps to frame more appropriate regulations;
➢ an increased legitimisation of both the policy process and the resulting regulatory package;
➢ an enhanced level of commitment and compliance on the part of user groups;
➢ lower transaction costs, with user groups internalising some of the costs associated with data collection, surveillance and control.

Some authors (Dubbink and van Vliet, 1995) also suggest that co-management implies a less legalistic approach to governance, allowing greater discretion to individuals to adapt their conduct to the spirit of public policy. But it is important to recognise that co-management does not necessarily mean a reduction in the level of regulation, but rather that those regulations are likely to be determined locally and, therefore, with greater sensitivity to the particular conditions of fishing activity. The individual fisherman does not become a free agent; under co-management, his actions are constrained in a socially controlled system by decisions that his peer group has helped to frame.

Co-management: Some Basic Questions

Although we have now moved closer to a definition of co-management and the expectations that may arise from its implementation, we are still some way from being able to describe the 'ideal structures' of co-management. The next step is to define the conditions under which co-management may prosper. To do this, we need to ask a number of questions concerning the *scope* of co-management, the *timing* of user group participation, the *representation* of user group interests and finally, the *geographical scale* of the co-management system.

Initially, it is possible to lay down some preconditions for co-management, which relate to the structures, capabilities and aspirations of the two parties involved: the state and the fishing industry. Above all else, co-management may be described as a 'state of mind', reflecting an unusual degree of mutual respect and trust for each other's aims, expertise and experience. The quality of the relationship will depend partly upon a willingness to develop greater flexibility in the approach to management, but also partly upon confidence in the organisational strength and integrity of the 'other' party. Here, perhaps, the greater challenge faces the organisation of the user group interests. The central government's administrative apparatus must be able to demonstrate that it is open to information and advice from external sources which may not be easy in circumstances where a science-led bureaucracy is used to hearing its opinions reviled by user groups. However, a common problem for the representation of user group interests is that they, all too often, reflect the internal divisions of a heterogeneous industry, divided along sectoral and/or geographical lines and whose capabilities and aspirations for deeper involvement in the co-management process may be limited.

A prerequisite for successful co-management is a relatively simple organisational structure incorporating strong, widely respected and fully representative user group organisations. According to Scott (1993), writing about self-governance in fisheries, the larger the organisation, the more difficult it is to maintain a democratic process based on direct participation. Thus, decision-making has to rely on an 'aggregation' of interests and compromise decisions, rather than 'integration' and a consensus based on 'communicative action' (Habermas, 1984). One of the problems for co-management is how to deal with the tendency for fission within fishermen's organisations creating smaller, more coherent interest groups. A proliferation of interest groups may lead to the isolation and alienation of some sectors whose interests are not formally represented within the co-management structure and may thus imply a threat to the perceived legitimacy of the policy process.

According to Hersoug and Rånes (1997), the range of functions assigned to co-management is commonly set too narrowly, the timing of the 'intervention' by user groups within the policy process is too late for maximum benefit and the spectrum of 'stakeholders' for inclusion within the co-management system unnecessarily limited. Typically, co-management is seen to be concerned solely with the organisation of harvesting activities and the stakeholders are thus confined to the harvesting sector alone.

Ideally, the management process should involve:
> setting policy objectives;
> resource estimation (stock assessment);
> structural measures to limit fishing capacity and modernise the fleet;
> other limitations on fishing effort, including the setting of TACs, quotas and effort restrictions;
> allocation of access to resources: who fishes what, where, and how?
> monitoring and enforcement;
> market regulation, including first hand sales and the regulation of exports and imports;
> infrastructural developments, including landing facilities, processing etc.;
> grant aid schemes;
> research and education;
> marine habitat and eco-system sustainability.

To a greater or lesser extent, all such activities should benefit from a co-management approach. That being the case, it is clear that user-group participation should be sought from the earliest stages and continue throughout the policy process and not be restricted, as is so often the case, to the implementation phases alone, by which time the decisions have, to all intents and purposes, already been finalised.

Much the most difficult problem for co-management, which reflects both the fundamental nature of user groups interests and the potentially wide range of co-management functions, is deciding which stakeholders should be represented and how those representatives should be chosen. As indicated above, most so-called co-management systems are based on single-interest representation drawn from the harvesting sector. But there are strong arguments to support multiple-interest representation in pursuit of a more broadly constructed agenda for co-management. Clearly, the transactions of the co-management organisation, however they may be defined, will benefit from the presence of specialist advisers (scientists, economists etc.). But the problem of representation is multi-dimensional: where to draw the line in terms of the extent to which downstream interests are represented (merchants, processors, consumer groups) and also laterally in terms of those who share an interest in the management of the seas. Expanding the range of interests, in either direction, threatens to create unwieldy structures and also increases the risks of conflicts within the co-management organisation. Moreover, striking a balance between competing user groups within the harvesting sector is not easy. Co-management organisations cannot afford to be too large; yet the alienation of particular groups risks the erosion of co-management's greatest benefit - the legitimacy of the system and its outputs.

The problem of representation may be compounded by decisions over how the membership of the co-management organisation is selected: as representatives of specific interest groups or as individuals, chosen for their knowledge and understanding of the industry but without a particular constituency base. The former raises doubts over the representatives' freedom to bargain or accept compromise solutions. But the latter process, designed to assist deliberations to rise above interest group politics, would seem to imply the selection of members by the state - and this may call into question the legitimacy of the co-management organisation. As such appointees will inevitably have some affiliation to particular sections of the industry or to particular geographical areas, they will find it hard to escape the pressures of the constituency - a situation made all the more difficult because they have not been elected to represent those interests.

In some respects, the solution to such dilemmas may be found in the answers to questions concerning the spatial scale of co-management. The simplest answer to the question 'at what geographical level should co-management operate?' is: at all levels from national through regional to local - but not in the sense that the system should recreate the conditions of hierarchical, top-down structures of management. Indeed, from an organisational point of view, a reversal of the structure would be more appropriate. The best opportunity for effective co-management occurs at the local scale; the surviving vestiges of self-management systems in European fisheries (*cofradía, prud'homie*) and the most durable examples of co-management (Sea Fisheries Committees in England and Wales) were developed at the local level. It is here that the remit for co-management is most likely to be restricted to the harvesting of the resource and for the 'balance of power' to rest with local user groups. Here, too, the opportunities for full representation of all sectoral interests - at least within the harvesting sector - are maximised. But co-management is equally important at the elusive regional level and at the national level; where the range of functions assigned to co-management organisations will be broadened and the range of stakeholders increased. By contrast, representation of the harvesting sector will need to be compressed and to be drawn from federal organisations, hopefully reducing the likelihood of internal conflicts and increasing the chances of consensus.

But there is a newly emerging facet within the co-management debate. According to Gislason (1994), 'the objectives of fisheries management are now gradually changing from the sustainable use of the commercially valuable resources to the conservation of marine environmental quality'. Although this is likely to have significant implications for fisheries management throughout areas like the North Sea, the challenge of environmental management will be most keenly felt in inshore waters and especially in areas where the marine habitat is valued not only for its commercial resources but also as the feeding ground for important wildlife species. Addressing the issues of integration of the sciences, resource user groups, environmental protection organisations and the state administration presents the toughest challenge to co-management.

Not only are new actors involved, but new conflicts may emerge. Sports and leisure fishing will demand recognition and, among the commercial interests, the numbers of part time, seasonal and partially retired fishermen will further complicate the problem of representation. The allocation of rights to these minority groups has long proved a problem for inshore management. Conflicts will arise not merely over the different aims and objectives of the resource user groups and the environmental interests but also over the interpretation of the science on which management decisions are to be made. In terms of 'sustainability', for example, what may be deemed an acceptable level of stocks for purposes of commercial exploitation may be judged inadequate in terms of eco-system sustainability. Science is still some way from providing an adequate operational definition of marine environmental quality and very little is, in fact, known of the long-term impacts of fishing on the marine eco-system. One consequence of scientific uncertainty is the likelihood that management strategies will be constructed on the basis of the precautionary principle - a certain recipe for dispute. Whatever the science, such is the power of the conservation lobby that the legitimation of the conservationist view, through public opinion, will almost invariably outweigh that of the fishing interests. With the inclusion of public opinion as a factor in co-management - hitherto lacking in fisheries management - the balance of power is newly disturbed.

Devolved Management Systems in Denmark, Spain and the UK

The focus so far in this chapter has been upon policy and institutional processes and more particularly on the theoretical considerations surrounding co-management systems. The remainder of this chapter attempts to assess co-management in practice through an exploration of new or existing opportunities for strengthening the role of fishermen's organisations and developing a more effective system of management for fisheries in the European Union (EU). Integral to this analysis will be a review of the role and positioning of fishermen's organisations in present and future management systems in three case study countries: Denmark, Spain and the UK. All have witnessed a significant growth of government intervention in the management of fisheries; the level of regulatory control has intensified, largely in response to a deepening resource crisis, but significantly at a time when in other sectors of the economy the prevailing trend has been towards deregulation and exposure to market forces. Under the growing influence of Brussels the bureaucratisation and continued centralisation of fisheries management has become a principal cause for complaint. With some justification, the harvest sector, together with its local knowledge, skills and expertise, claims to be progressively marginalised and discredited in the decision-making processes.

A distinction needs to be drawn between policy formulation, located within the EU's policymaking institutions, and implementation which remains the responsibility of the individual member states. Thus, although the broad strategy for fisheries management is defined in Brussels, along with many of the key parameters for its legislation, there remains considerable scope for variation in the detailed implementation of fisheries policy at the level of the nation state (Symes, 1995). This encourages the development of a wide range of detailed policy processes, structures and outcomes within the community of fishing nations in Europe. The division of responsibility also implies that opportunities for the devolution of management responsibilities will rest largely at the level of the member state and are likely to vary in form. Hence the focus on national cases within this chapter.

Of central importance to the emergence of diverse management approaches and opportunities for devolved management are the prevailing political and institutional cultures of the fishing industries and their internal political organisation. *Institutional cultures* and decision-making approaches are divergent in form and this affects the degree to which user groups are incorporated within the management system. In summary, Denmark, with a centralised political system, has a well established tradition of engaging the participation of interest groups within formalised procedures for consultation over policy development, in a system commonly referred to as a system of 'centralised consultation'. By contrast, the UK retains a strongly centralised bureaucratic system of administration and policy delivery, but with little regular and formal consultation between the state and the user groups on matters of detailed policy. This is despite the fact that fisheries administration is divided between different 'provincial' departments in England and Wales, Scotland and Northern Ireland. Meanwhile, Spain has evolved as a 'regionalised state', with responsibilities for economic and social policy decentralised to the Autonomous Communities. Nonetheless, in Spain resource users are relatively marginalised from the decision-making process.

The pattern of *political organisation* within the industry is clearly essential in determining the scope for and design of user participation systems. Throughout the three case studies the fishing industries represent a very loosely structured, unstable and often discordant coalition of interests which reflect not only differences between the harvesting and post-harvest sectors but also their fragmentation according to particular sectoral, structural and regional preferences. This cleavage of interests is reflected in a fragmentation of political representation of the industry into regionally and sectorally based organisations, usually co-ordinated through national federations, though rarely do these include all the relevant groupings of interests. Fission rather than fusion tends to characterise the evolving structure of political representation and there can be little doubt that such weaknesses have contributed to the marginalisation of the industry's influence within the institutional system. There appears to be a potentially significant relationship between the general political culture and the organisational structure of the fishing industry. Where the industry is given a place at the negotiating table, as in Denmark, there is a tendency for stronger and more coherent representative organisations to emerge. Where, on the other hand, representative organisations compete for the attention of government departments through informal consultation, fragmentation and conflict of opinion may be frequent.

This case study analysis is divided into two sections. In the first the form, function and structure of fishermen's organisations are considered. This involves a broad analysis of the existing level of user participation with regard to policy formulation and implementation. Some attention is then paid to notions of opportunities and alternative institutional systems for the countries under analysis as well as the European Union as a whole. Following this, some conclusions are drawn concerning devolved management systems and co-governance.

Fishermen's Organisations: Structure, Function and Role within the Policy System

A baseline analysis of fishermen's organisations (FOs) is essential in order to appreciate the viability and location of existing or potential loci of user participation. FOs demonstrate a potentially confusing diversity of forms in terms of membership size and characteristics and in their level of geographical identity and formality. Important distinctions can be made between those organisations that reflect sectoral (e.g. employers v labour; large vessel v small boat; pelagic v demersal) or regional interests and those which are statutory or non-statutory in form. There is also diversity in terms of whether membership is all-embracing (i.e. vessel owners and crew members) or restricted to certain categories of persons. Few FOs are comprehensive in their membership, in the sense that they comprise all active fishermen in the port or district. Some extend their rules of membership to admit other local interest groups located downstream from the harvesting sector (e.g. salesmen, merchants, local processors, etc.). Despite their origin under a range of legislative and institutional systems, most FOs are in fact co-operative rather than corporate structures. FOs range from the smallest port-based associations to the larger regional and national organisations; such separate levels of representation are often integrated through a hierarchically structured federal system.

A key division in terms of function can be drawn between those FOs which are dedicated to political representation and those concerned with management functions in respect of both production and marketing. This chapter considers the role of two key forms of FO within the harvesting sector notably fishermen's associations which principally represent the welfare, management and commercial interests of the harvesting sector, and producers' organisations which were originally developed for the purpose of organising the markets for their members' catches. Both organisational forms are democratically based, operating through consensual decision-making in most instances, and self-governing.

Fishermen's Associations (FAs) aim to represent the interests of their members in negotiations with policy makers either through direct negotiation, through lobbying of civil servants, parliamentary representatives and government ministers, or through representation on formal consultative or advisory committees. Representation occurs at different geographical scales - national, regional and local - and thus the negotiation may be with central, regional or local authorities. Equally important is the process whereby FAs relay information and advice from the authorities to their members whether as individuals or constituent organisations. The FA remit may extend well beyond this political role to include the provision of legal and financial services and to a range of other commercial activities and, less commonly, to the regulation of fishing activities.

Fulfilment of the FAs' political objectives usually calls for vertical structures linking the local grass-roots associations with the larger generally national organisations. A distinction can be made between those organisations which are *'federated upwards'* in the sense that local associations are built up into larger regional and/or national organisations and, secondly, those which are *'devolved downwards'*, where a centralised organisation is divided into regional and/or local groupings in order to tap into the strength of grass-roots opinion. In almost all cases within the fishing industry, the former prevails. Indeed, all three countries conform to this hierarchical pattern of organisation with local associations at port or district level, together with regional and/or national federations. It should also be noted that federation does not necessarily mean greater political influence. There are several instances where federations are weaker or less active than their constituent associations.

The basic functions and form of *producers' organisations* (POs) are broadly similar throughout the EU. As EC inspired organisations, their primary role since the initiation of the Common Fisheries Policy, has been in the common organisation of the market for fisheries products, notably through the operation of the withdrawal price mechanism (Hatcher and Cunningham, 1994). Their purpose has been to ensure that fishing is carried out along rational lines, providing maximum returns for their members' catches - hence their function is commercial in orientation. While initially established to help bring order to local first hand sales of fish, more recently in the UK their functions have been extended to include aspects of quota management. This provides a good example of the diversity in management approaches as a result of the division of responsibility between Brussels and member states. Despite similarities of broad function and format, POs display considerable diversity in terms of membership characteristics (especially size), geographical identity and even remit. POs are voluntary, mainly co-operatively based organisations and mostly financed by means of a levy on members' landings. Membership is sought on either a company or an individual vessel basis and spatial patterns of membership display considerable overlap. There are also significant areas not covered by POs, particularly in areas dominated by small scale, traditional inshore fisheries.

Attention now turns to the structure and function of FOs in each of the case study countries under analysis. This also involves consideration of the various institutional contexts involved and, more particularly, the degree and location of user participation within the policy system. It will be seen how differences in organisational patterns and styles of governance clearly reflect diverse political, cultural and democratic traditions. The level of user group participation in systems of policy formulation and implementation is, in fact, an embodiment of the political culture of fisheries management.

Denmark

Scandinavian countries have a tradition of facilitating the participation of interest groups in policy discussions concerning the determination of objectives for specific sectors of the economy. This is equally true for the fishing industry and the only example of a formalised system of consultation among the three case study countries is found in Denmark. In part this situation may be linked to a more rationalised structure of fishermen's organisations. As a small country with an extensive fishing industry, Denmark is distinguished by an outwardly simple structure of organisation, though this may conceal inner problems of consensus building. Indeed, the Danish Fishermen's Association (*Danmarks Fiskeriforening*, DF), representing up to 80% of Danish fishermen through its affiliated organisations, is the result of an amalgamation in 1994 of two former regional associations. Hence, decision-making within the organisation involves a complicated process of compromise between the regions and is politically sensitive. Like many federal organisations, it faces a number of problems of balance - as between the different regional components of the industry, heightened by the concentration of its membership in the north, and more especially in its role as an employers' association in wage negotiating, while claiming to represent a cross section of interests throughout the industry including both vessel owners and crew members alike.

Danish FOs have an important consultative role in respect of structural measures and the regulation of fishing effort through the *Regulation Advisory Board* (with a broad representation including harvest, processing, exporter, research and Ministry interests). This Board advises the Fisheries Ministry in matters relating to fisheries regulation including the allocation of quota. The *EU Advisory Board*, again with multi-interest representation, provides a further link, advising the European Committee of the Danish parliament on matters relating to the development of EU policy. Both boards are advisory and do not confer any executive responsibility or exert undue political influence on the FOs. DF representatives have also worked with representatives from the Research Directorate and the Ministry on two recent working groups considering an alternative effort limitation package and the broader issues of CFP reform.

In contrast to Spain and the UK the number of POs in Denmark is small and stable. Only three POs have been established; one national, one sectoral and one with a strong regional identity. The Danish Fishermen's PO (*Danske Fiskeres Producentorganisation* or DFPO) has a virtually nation-wide membership and contains almost the entire demersal sector and therefore its membership naturally overlaps that of the DF. In range of functions, it is the most restricted, concerned primarily with the implementation of minimum prices and quality controls as initially laid down by the EC. The second demersal PO, Skagen FPO (*Skagenfiskernes Producentorganisation*) based in North Jutland is also involved in the organisation of sales contracts. By contrast, the pelagic Purse Seine PO (*Notfiskernes Producentorganisation*) has developed a much wider range of responsibilities including the planning of the fishing activities for the pelagic fleet, the management of herring and mackerel quotas, and external negotiations with Norwegian and Swedish fishermen's organisations to manage supplies of herring.

There are additional *ad hoc* experiences of user participation in Denmark (Vedsmand, Friis and Raakjaer Nielsen, 1995). For example, as in the other states under analysis, there are instances where local government has harnessed FOs within regional initiatives for the development of fisheries dependent areas. Notable examples in Denmark have taken place in Bornholm and North Jutland each involving a multiplicity of interests from local government, catching and processing sectors. There are also other regionally specific examples of user participation in Denmark including the FO-based Matjes Committee, which manages the Matjes herring fishery on a basis of weekly catch rations in the North Sea and Skagerrak.

Spain

In *Spain* the governing situation is complicated by a division of responsibilities for fisheries management between the central administration and the Autonomous Communities. The particular division of responsibility is certainly disputed; currently the Autonomous Communities ascribe responsibilities for 'interior' waters (those waters lying between the coast and base lines) within a frame of reference established by the state, which has control over the 'external' waters out to the twelve mile limits (Alegret, 1996). Industry representation is absent from the *Junta Nacional Asesora de Cultivos Marinos* and the *Junta Asesora de Pesca Maritima* which form the bridging organisations between the two administrative levels. There is little or no formal liaison, at either national or regional level, between representatives of the industry and the administration although there are cases of particular consultative efforts on behalf of particular administrations; in Galicia, for example, there is a formal consultative body (*Consejo Gallego de Pesca*) in which the FOs and regional administration are jointly involved. Informal consultations take place between central government and the national committees of the industry's representative organisations.

There has been little in the way of delegation of resource management functions to Spanish fishermen's organisations. There are four main categories of FO in Spain including *cofradías* and Vessel Owners Associations, trade unions and the EU instigated POs. The port based *cofradías* (fishermen's guilds) are derived from powerful medieval guilds. They exhibit surprisingly formal structures with democratically elected Permanent Commissions. The Spanish *cofradías* are grouped into regional federations that combine into a national federation structure. The regional level of federations is generally weak, but serves as a medium for communication between the *cofradías* and regional administrations. The National Federation of Fishermen's Guilds provides the only definite links to MAPA (central fisheries administration) and the European Union advisory bodies.

The activities of *cofradías* are commercial (control of local markets) and social (provision of welfare services) in orientation. Their resource management functions are more limited and concerned primarily with the regulation of fishing activity for the benefit of the market and the local resource base. The Spanish *cofradía* is a broadly constructed, elective fishermen's organisation (with equal representation of vessel owners and crew members on the Executive Committees) with obligatory membership for all actively engaged in the harvesting sector, reflecting the *cofradía's* monopoly position in the handling of first hand sales of fish on the quay. It exercises certain responsibilities, jointly with the central and Autonomous Community administrations, for the regulation of fishing activity. In effect, the *cofradía* manages the local fishing industry, rather than the resource base, through the delimitation of fishing areas and seasons and the establishment of timetables for the departure and return of fishing vessels and the hours of opening for the local auction markets. In theory, each *cofradía* will establish a 'nuclear' territory, based on proximity to the home port (Alegret, 1996) though in practice the fishing grounds of neighbouring *cofradías* will tend to overlap. On the rare occasions where the *cofradía* imposes maximum daily limits on members' landings, this is intended to achieve equity and a balanced market, rather than as a resource management measure. As the *cofradía* also has the capacity to determine who may or may not be admitted to its membership and as membership of a *cofradía* is in effect a condition of being granted a licence to fish, the *cofradía* may also be seen to possess rights, conjointly with the central and regional administration, to exclude fishermen from access to the fisheries.

The membership of vessel owners' associations overlaps that of the *cofradías* and attracts those from among the large scale sector of the fleet. Although weaker than the *cofradías* counterpart - experiencing greater difficulty in organising members as a cohesive unit - the vessel owners' associations also form a federal structure which is also involved in discussions at MAPA and EU levels and is particularly interested in negotiations with third countries, reflecting its predominantly deep sea interests.

The creation of the 45 POs in *Spain* has been quite controversial: they are seen to compete with the well established *cofradías*, which have traditionally provided quayside services and organised the auctions for the members' catches. Replacement of the *cofradías* by POs would clearly be seen to peripheralise active crewmen as they would be excluded from both PO and Vessel Owners Association membership and so be confined to organisation within Trade Unions. Indeed, the absence of co-operation from *cofradías* has slowed the pace of PO development. The majority have been formed through co-operation between neighbouring vessel owners' associations as the functions of the two organisational types often appear to overlap.

United Kingdom

The UK organisational and decision-making structure is complex. Its system of governing is divided regionally between provincial government departments which helps to provide a more regionally sensitive governing system. However, there are debatable levels of autonomy within the provinces and the system remains essentially bureaucratic and centralised within the Ministry of Agriculture, Fisheries and Food (MAFF); MAFF retains lead department status with responsibility for UK fisheries as a whole, in addition to its provincial remit in England and Wales. The division of responsibility does introduce several potential lobbying points for the provincial industries. There are drawbacks however, as highlighted in Scotland; while the Scottish Office is arguably more open to consultations and opinions from various Scottish FOs, when compared to the situation south of the border, some are critical that they are in any case effectively distanced from key decision-making in Whitehall.

As a whole, industry opinion is not seen to be integral to the policy formulation process within the UK. Currently there is no formal or transparent system of consultation on the broader aspects of policy and informal, *ad hoc* discussions are the only forms of dialogue. A complicated organisational structure may be partly to blame in terms of the 'provincial' department arrangements. But perhaps a more important factor is the degree of fragmentation within the representation structure within the UK through the local associations and their federations.

Varying in size, vitality and internal organisation, the proliferation of local port based *fishermen's associations* provides a framework for collective representation and a forum for the discussion of local issues. They are mainly associated with inshore fisheries and are therefore largely made up of skipper-owners. Some are, in effect, residual - almost relic organisations - reliant upon a very small active membership and financially weak. Most are much stronger and politically active, and provide a range of commercial services. Reflecting, in part, the separation of responsibility for fisheries administration between the provincial government departments, a dual federation structure exists with the National Federation of Fishermen's Organisations (NFFO) representing interests in England, Wales and Northern Ireland (through the numerous port based FAs) and the Scottish Fishermen's Federation (SFF) incorporating eight regional and sectoral associations across Scotland. A large number of independent local associations and regional federations occur outside the national federations for various reasons (policy, finance, personality etc.). As a whole, the pattern is one of considerable fragmentation of interest groups which poses considerable problems for the industry in promoting a united view. There have been limited attempts to rationalise the organisational structure; for example, in order to improve liaison with local and regional organisations and secure a more stable financial base formal links have been created in England and Wales between the NFFO and certain producers' organisations with seats established on the Federation's executive committee.

In contrast to the situation concerning policy formulation, regular consultations with user groups do emerge over detailed aspects of policy implementation where FOs have a key role. Specific responsibilities have been delegated to producers' organisations and Sea Fisheries Committees.

Producers' organisations (POs) within the UK have, in fact, displayed a progressively expanded remit. Responsibilities for quota management have assumed primary importance among their functions; some have also developed their commercial activities in terms of marketing and processing initiatives. Their quota management responsibilities are administrative, involving the allocative task of assigning quotas to member vessels. A PO has no powers to prevent movement of vessels (and their share of quota) out of membership and into other POs; hence quota rights are not vested in the PO as such. Within most of the POs there has also been a growing involvement in the political arena and the distinction in terms of function from Fishermen's Associations is increasingly less clear. Substantial geographical overlaps occur in the membership of the POs and their number has increased to 19, often as a result of defection from *or* disaffection with already established POs. Some are federated into a national organisation although its role has been somewhat ill defined.

At the local level, though only in England and Wales, users also participate in policy implementation through the regionally based *Sea Fisheries Committees* (SFCs). SFCs have a wide range of powers to regulate fisheries within the exclusive territorial waters to a distance of 6nm from the shore. They are non-elective organisations, financed by local authorities, comprising appointed representation from local councils and the fishing industry in equal numbers, with the latter appointed by the central government department (MAFF). Essentially, the 12 SFCs exercise responsibility for the regulation of inshore fisheries through a series of legal instruments including bylaws and Several and Regulating Orders; the latter apply only to the mollusc^2 fisheries and confer specific property rights on individual fishermen and licensing powers to the SFCs. By-law regulations provide for the introduction of ground closures, gear restrictions, access limitations (based on vessel size) and the imposition of minimum landing sizes and, less commonly, quotas. A key feature of the SFCs remit is their added responsibility for the policing of all fishing activities within their areas of jurisdiction. SFCs also serve to link concern for the sustainability of the inshore fisheries with the conservation of the marine environment. As a whole SFCs, dedicated solely to the management of inshore waters and acting as quasi-autonomous organisations in respect of their prosecuted fisheries districts, may be seen to exercise the strongest degree of devolved management at the local level in all the cases under study and arguably throughout the European Union.

Comparison of Cases

Clear differences are evident in the way in which FOs are organised within the states under analysis. The main contrast is between Denmark, where the structures have attempted to capitalise upon 'economies of scale' and increase their external influence and internal control through the creation of fewer and larger organisations, and the UK and Spain, where there is a continuing fragmentation of representation and proliferation of management organisations. The two main types of organisation present, FAs and POs, also differ in terms of the location of their strengths. FAs tend to be relatively weak in organisation at the local level but gather strength through federation. POs, by contrast, exert their greatest influence at the local or regional level but lack strength of organisation at the federated level. The political strength of the national associations derives from the grass-roots representation of their local or regional associations. It is, however, relatively rare for a national fishing industry with distinct cleavages of interest to be represented by a single, overarching organisation. Problems may occur where a federation is unable to resolve strong differences of regional or sectoral interest from within its membership, which may lead to a splintering of the federation, and the negotiating position of the industry as a whole is weakened. Whilst vertical co-ordination of FOs is often strong through federalisation, horizontal co-ordination between different forms of FO at both national and local levels is usually weak. Sometimes it is only achieved through informal mechanisms such as the overlapping of membership between FAs and POs at board level, and less commonly through the sharing of executive responsibilities and administrative services.

In passing, and notably in the case of producer's organisations in the UK, it has been noted how in some cases FOs have a clear role in the implementation of policy measures. In most instances this involvement is essentially concerned with administrative tasks rather than more fundamental delegation of autonomous authority. Nowhere has the responsibility for fisheries policy been transferred from 'central' government to a delegated, non-government institution; there are, therefore, no instances of fisheries self-management. In some of the cases, there are also devolved local systems for the regulation of fisheries focusing principally on inshore fisheries. Some are deeply rooted in historic institutions (e.g. the *cofradía* in Spain) while others have been brought into being more recently as part of the formal system of fisheries administration (e.g. Sea Fisheries Committees in the UK).

The extent of incorporation of the knowledge, experience and understanding of the fishing industry within the policy making system also varies between the study nations. A clear division can be made between the 'negotiation economy' of Denmark, where policy issues are debated fully within a formalised system of consultation, and the more bureaucratic systems in the UK and Spain, where there is an absence of formal machinery for negotiation and consultation arrangements are typically *ad hoc*, informal and less transparent. It is, in fact, difficult to judge whether consultative arrangements, even when well established, do in fact materially affect the direction of policy. If anything they create a sense of engagement which, might, at the end of the day, be their sole intention. All the systems of consultation are tested and challenged on their ability to reach consensus views. It is important also to remember that the framing of policy is largely undertaken by the European Commission and decided upon by the Council of Ministers. National policy making is therefore largely limited to the interpretation and implementation of EU regulations or to areas where the member state has unqualified discretion. As a result, the fishing industries in all three countries are one degree removed from the central decision-making arena. They can, nonetheless, bring their influence to bear on Ministers, who represent national interests in Council meetings. In addition, Europêche as the federal organisation representing the national associations of fishermen throughout the EU can seek to influence policy deliberations through its representatives on the Advisory Committee for Fisheries.

It would appear from the foregoing analysis that the level of direct involvement of FOs within the formulation and implementation of fisheries management policies remains relatively low. In Denmark, FOs are admitted into formal discussions over fisheries policy but there has been a reluctance on the part of government and FOs alike to become directly involved in managing the outcome of these discussions in the form of devolved quota management. In the UK, the virtual exclusion of professional organisations from formal consultations with central government departments over policy determination contrasts with the well developed sectoral quota management system which devolves responsibility for the routine administration of quotas to producers' organisations and the regulation of inshore fisheries through Sea Fisheries Committees. Meanwhile, in Spain the traditional local structures with wide ranging involvement in both economic and social aspects of policy implementation - the *cofradías* - are presently under siege from the attempted imposition of POs.

New Opportunities for User Participation

The current state of user participation in the three countries is clearly divergent; this reflects particular institutional and democratic traditions and various strategic and structural constraints. These characteristics will ultimately condition the choice of new opportunities and structures for devolved management.

Limits to Participation

An awareness of organisational capabilities, both structural (size, membership and administrative resources) and strategic (motivation and aspiration), can help in understanding the existing or potential position of FOs within the policy system. A lack of *strategic capabilities* can be of central relevance to the absence of user participation in decision-making. In many local and regional organisations there are few aspirations for an extended remit in policy framing and implementation. There are exceptions to this situation such as in North Jutland, where a case is being put forward for the co-management of local quotas through a partnership of local associations and the processing sector, and in the Highlands and Islands of Scotland, where some local associations aspire to a more proactive role in inshore fisheries management in partnership with regional authorities. As a whole, however, conservative, defensive attitudes and reactive modes of response tend to predominate over innovative, progressive and proactive strategies for development, at several levels of organisation within the fishing industry. Inertial forces within government and within key industry organisations often stem from mutual suspicion and a lack of conviction that institutional reform will bring sufficient tangible benefits; this may simply be conditioned by years of deteriorating relations caused by the persistent failure of policy measures to achieve the necessary turn round in the health of the fish stocks. In some cases the industry has failed to demonstrate an awareness of the dynamics of interaction with government and the policymaking process in general.

Concepts of devolved management do not sit comfortably within the prevailing political cultures of the three countries. Fundamental reform of the institutional arrangements for fisheries management, particularly in terms of devolved management, is not high on the political agendas of either the state or the fishing industry. There is considerable opposition to notions of change from 'the establishment', that is from both central government and some of the fishermen's organisations located at a relatively high level in the hierarchy of the present system.

Structural elements of FOs also function as limiting factors and, in part, a lack of motivation for an expanded remit may be attributable to an awareness of these features. There is clearly a problem of co-ordination and co-operation; diversity and conflict pose significant barriers to the successful devolution of responsibility. While smaller FOs may have a clearer identity of purpose and coherence of interests, larger, federal organisations representing a very broad constituency of interests, must attempt to reach a consensus in decision-making. Apparently stable coalitions of interests can be threatened by a single major issue and internal conflicts can considerably slow down the internal decision-making process within FOs. A perceived inability to represent certain interests in negotiation, or a sense of power imbalance within an organisation, can deepen internal cleavages and create fission. Fracturing is not uncommon among the larger FAs and federations and this places some uncertainty over the representative capacities of FOs in general. There is considerable difficulty of representing rank and file members and developing effective systems of internal communication through an often hierarchical structure of local and regional branches. As a whole, confrontation is frequent within the industry, which may experience conflicts of aims and objectives between user groups employing different gears or between different stages in the food system (harvesting, merchanting, processing). Where the attempt is made to combine a broad spectrum of interests within a system of co-governance, there is concern that there may be paralysis of strategic decision-making and weak, consensual compromise; and this may deter closer relations between state and user group.

For most FOs there are also financial constraints. In smaller associations this may translate into personnel difficulties and a dependence on low paid or voluntary part time assistance from former fishermen or other shore based persons. In part this is typical of a wider recruitment problem for smaller organisations where decision-making responsibilities are often devolved to a handful of willing persons or dominant members - participation by active fishermen is often constrained by the unsociable working hours and by a high degree of apathy. Larger, financially stronger regional associations and federations will usually have the advantage of a regular administrative staff and professional leadership. Financial constraints may also restrict the range and strength of commercial activities carried out by an FO.

Producers' organisations display generally sounder financial and structural capabilities, although financial limitations may also impose a constraint upon their activities. Organisational strength is attributed by their strategic role in the organisation of the harvesting sector and their degree of knowledge and influence at the regional level. Ironically, whereas POs are in much closer and more regular contact with the conditions and issues of the local fisheries than most FAs, they have less formal opportunity for expression of views at the national level. POs face a number of specific structural problems that might deter support for their candidature for a central role in devolved management:

> ➤ *overlapping memberships, boundaries and functions*; this is particularly the case in Spain where there is confrontation with the *cofradias*, a slow and contested development of POs, an incomplete national coverage and a duplication of functions; in the UK, PO numbers have spiralled and with overlapping boundaries there is often the confusing situation of several management regimes operational in a single locality;

> ➤ doubts over their ability to take *effective disciplinary action* when in-house rules are broken; this applies particularly to the system of sectoral quota management in the UK, where rumour rather than evidence of poor discipline is common; enforcement of rules is prejudiced by the freedom of movement of members within a voluntary system of membership and the risk of losing a member vessel's quota record;

> ➤ a failure to achieve *effective supply management* or to exploit their potential for effective, pro-active management; few have extended their remit to secure orderly marketing conditions for their members by developing interests in the processing sector; part of the problem relates to the adequacy of the existing financial and administrative infrastructures rather than willingness or competence.

From the analysis so far, it is clear that in all three countries, the structures currently in place are more or less well suited to the particular tasks of representation and a limited range of devolved management responsibilities. Not all the constituent organisations share the same level of capability in terms of finance, leadership, management skills and aspirations. Indeed, the analysis has indicated that many small, local organisations are unable to adequately fulfil their responsibilities for want of sufficient financial and administrative resources.

While the industry is indeed overpopulated in terms of numbers of organisations, it is argued that if these local organisations were to disappear through natural wastage or amalgamations, significant losses might occur in

Fig. 2: A Capability-based Typology of Fishermen's Organisations

Type	Summary	Fishermen's Organisations
High Structural Capability High Strategic Capability	Maximise both existing strengths of organisation and opportunities for further development. Tempered by a degree of realism. Often display a degree of flexibility that has enabled them to reposition themselves in a changing structural and policy environment.	Most of the federations of FAs. Some – but relatively few - of the larger local or regional associations and POs (notably those which exhibit sufficient knowledge of the markets and the ability to manipulate supply).
High Structural Capability Low Strategic Capability	Competent but undynamic leadership. Lack the motivation for enhanced roles but have the structural characteristics to enable them to carry out their present range of activities adequately. Sufficient flexibility of structure to cope effectively with changes in environment. Possible reasons for current lack of aspiration may include: personal characteristics, an underestimation of structural capabilities, statutory limitations.	Majority of POs and many of the medium sized and larger FAs.
Low Structural Capability High Strategic Capability	High aspirational goals but lack the power of numbers and adequate financial support to make a significant impact.	Relatively uncommon situation often affecting those FOs not affiliated to federation.
Low Structural Capability Low Strategic Capability	Little by way of decision-making procedures and with inadequate financial and administrative back up. They often survive through the efforts of individuals or small groups. Low aspirations, conditioned by a high degree of scepticism. Effective at providing local grass-roots representation.	Mostly informal and small associations.

terms of mobilisation of community action, grass-roots opinion and the democratic basis of the organisational structure. Nonetheless, all three countries have a hierarchical system capable of delivering grass-roots opinion to the central decision-making institution, albeit in both Spain and the UK the final link in the chain of information and advice is missing. Likewise organisations with responsibilities for a limited range of devolved management functions, mainly in the field of first hand sales, occur in all the case countries. Figure 2 (above) considers a typology of fishermen's organisations based upon an analysis of structural and strategic capabilities.

Opportunities for Participation

It is the natural outcome of addressing institutional constraints that proposals for change are not overly radical. They will likely be based essentially on the realignment and relocation of *existing structures* and the channels of communication that link them. They could not conceivably challenge the leading roles of either the European Commission or the nation state in framing the broad principles of fisheries management policy and the detailed implementation of surveillance, monitoring and control procedures; nor could they usurp the principal functions of the main organisations within the fishing industry.

From such a pragmatic perspective some opportunities do present themselves. In most of the case studies the institutional principles underpinning these are relatively similar, despite differences in the detailed means of operationalisation. They include:

➤ greater co-operation among fishermen's organisations and between fishermen's organisations and government;

➤ a strengthening of the strategic and structural capacities of user group organisations; fishermen's organisations need to *professionalise* their activities through a greater awareness of the dynamics of the interactions within the global economy and within the global, regional and national policy making process and the development of improved inter- and intra-organisational communication; they should also continue to explore the means for further rationalisation of their structures in a search for the further economies of scale and an increase in their ability to influence the conduct of the fishing industry;

➤ the transfer of additional responsibilities (quota or effort management) to appropriate fishermen's organisations such as POs; this has already proved successful in the case of the UK;

➤ the need for greater transparency and formality in consultation procedures.

In general terms, new institutional designs need not be complicated. It is clear, however, that they must be tailored to the particularities of the national context in different member states. For example, elsewhere there have been proposals for a integrated co-management structure for the *UK* in which existing organisations are restructured and realigned within a new institutional framework (Symes and Phillipson, 1996, Phillipson and Crean, 1997). The proposal for a series of Advisory Panels (APs), structured along ICES (International Council for the Exploration of the Sea) Area boundaries, is intended to provide a basis for more meaningful, transparent and formal consultation between central government departments and the fishing industry. Membership would be broadly structured, with representatives from central government, harvest and post-harvest sectors and from the statutory environmental organisations. To complement these, eight Area Management Committees (AMCs) are proposed to provide a more structured form of co-ordination of the existing delegated management groups (POs and Sea Fisheries Committees) and resource user interests at a regional level. Smaller in size and with their membership overlapping those of the APs but focused on the harvest sector, the AMCs would also provide a channel for grass-roots concerns to reach the national arena. Some opportunity for strengthened local/regional management of inshore waters through SFCs (England and Wales) are identified and producers' organisations are encouraged consolidate their disciplinary procedures and to maximise their current commercial remit with regard to marketing.

In the context of *Denmark*, Raakjaer Nielsen, Vedsmand and Friis (1997) have outlined a potential new institutional framework. Their alternative models distinguish two broad types: multiple-interest agencies and simpler, more narrowly defined, single-interest organisations; for each type there is a choice between national and regional institutions; this reflects the strong democratic tradition in Denmark where the norm is for all interest groups to be represented.

Thus, the proposals include:

➤ an Executive Management Council, with a broad membership drawn from all sectors of the fishery-related economy, which would transform the functions of the existing Advisory Boards into executive authorities with powers to act on a wide range of management responsibilities;

➤ Regional Management Councils, a regionalised form of the above with a much more restricted range of responsibilities, concerned primarily with the regulation of specific fisheries;

➢ a National Fishermen's Management Council, distinguished from the first model by its narrower membership confined largely to fishermen's organisations and which would transform the role of the existing Danish Fishermen's Association and/or the national PO to a series of executive functions related to structural planning of the fishing fleet and the formulation and implementation of regulatory measures; and, finally

➢ Regional Fishermen's Management Councils, where decision-making authority is located at the regional level and responsibilities are devolved from the fisheries department and its advisory institutions to the regional or local level.

Proposals can also be devised for *Spain*. Suarez and Frieyro de Lara (1997) have generated specific management models for particular regions of Spain tailoring the detail to the particular institutional culture of different Autonomous Communities. The proposed system is therefore sensitive to the recently developed and continuously evolving system of regionalisation of the political system. It reflects the considerable diversity at the regional level in terms of geographical location, stage of development in fisheries management, and the particular nature and structure of the regional fishing industries. For example, in line with the spectrum of alternative models outlined at the outset of this chapter, decentralisation of responsibilities to the Autonomous Community administration is considered most appropriate in the particular context of the remote location and fragmented structure of the Canary Islands; delegation of responsibilities (e.g. fishing plans, marketing and regulation tasks) to existing fishermen's organisations (the *cofradias*, vessel owners associations and producers' organisations) is preferred in the case of Andalusia; and the concept of co-management, in the form of a Fisheries Council, is presented as the most suitable model for the more sophisticated political economy of fisheries in Galicia (see also Chapter 11).

  Clearly, devolved management systems can be partial and disaggregated at a range of levels, among different institutions and to different extents, depending on the task at hand. This introduces a wide variation of potential designs for devolved management systems. There can be variation in the degree of devolution of management tasks in the sense of consultation, shared responsibility or full delegation. Even where tasks are best located centrally (enforcement, setting of TACs), there are still opportunities for the involvement of fishermen's organisations, especially in the input of the fishermen's knowledge and experience in relation to stock assessment. Other tasks such as the setting of technical regulations can be shared at the formulation stage and fishermen's organisations have a clear role in the implementation of responsibilities like quota management. A distinction emerges between 'policy functions', which are primarily retained by government, and 'operational functions' which have a greater potentiality for transfer to devolved agencies.

The immediate opportunities for devolved management in the three countries would tend to gravitate, to varying extents and with various combinations, to the less extreme forms of devolved fisheries management (delegation, decentralisation and co-management). The *enlightened dictatorship* variant is dismissed since it embodies the kind of technocratic and bureaucratic systems of management which many hold responsible for the failure of management systems today; arguably many of the existing institutional approaches would tend to resemble this variant when current instances of user participation are primarily centrally determined and confined in extent. Similarly, the prevailing political cultures are unlikely to tolerate any move towards *self-management*. There is the latent fear of anarchy within fisheries management should there be no central co-ordinating discipline and a concern for apparent deficiencies in the capabilities of existing organisations.

The predominant locus for devolved management rests at the point of policy implementation, within the jurisdiction of the nation state, and in the direction of co-management. In all three cases, opportunities would appear to lie in the realignment and consolidation of existing organisations and linkages within the policy system to allow more meaningful co-ordination, co-operation and consultation, and to a lesser extent in the allocation of new responsibilities to fishermen's organisations. In terms of policy formulation, possibilities are located in the consultative arena. The concept of a Fisheries Advisory Council is applicable to all the cases, though variously located at national or regional level and exercising either executive or consultative functions. In all instances the perception of the actors involved is seen to be crucial: if advisory arenas are seen as talking shops rather than more meaningful and fundamental mechanisms for the incorporation of user groups, then they may quickly disintegrate, except perhaps where they are seen by the participants as providing a very real opportunity to enhance their profiles and political prestige.

At the European level, however, there may also be room for institutional development. A regionalised co-management system has been proposed for the EU's fisheries (Symes, forthcoming) in order to introduce greater regional awareness in policy formulation and further satisfy the requirements of subsidiarity. In accordance with the principle of subsidiarity, the present monolithic Common Fisheries Policy would be replaced by a series of semi-autonomous regional Commissions, based on ICES Areas and made up of fisheries administrators and user group representatives drawn from the relevant coastal states and those with active fishing interests in the area. The regional Commission would determine all aspects of fisheries policy, subject to common principles determined in Brussels; it would also provide a rare example of structural coherence throughout the EU. In the case of the North Sea region, for example, the Commission would comprise the six EU coastal states, together with Sweden admitted on the basis of active fishing interests. Norway would be granted observer status at the meetings of the Commission. Ideally, such a system would build up from the national (and regional) Fisheries Councils. This would help to ensure a bottom-upwards flow of information, assessment and advice from the local and regional levels to inform the CFP as a whole.

Conclusion

The emphasis in this chapter has been placed upon the approach to governance in European fisheries. It has argued for a realignment in the patterns of interaction between regulators and the regulated through devolution, and more particularly co-management. Hence the discussion has transgressed between two levels of analysis (Kooiman, this volume), second order governing, through a consideration of alternative forms of devolved management, and meta-governing, in terms of the overriding and determining political and institutional framework. Consideration has been given to some potential opportunities for a shift in the mode of governing in three EU states and the European Union as a whole.

In the current state of Europe's fisheries, co-management is not an option; it is a necessity. The pooling of resources and expertise in a co-operative system can, in theory, only serve to enhance the quality of management. It may be the most opportune mechanism for coping with the highly complex, dynamic and diverse set of actors and issues within the socio-political context of fisheries governance. However, institutional change, the shift from one governing regime to another, tends to occur incrementally and to build upon existing organisations rather than result from the 'big bang' of radical reform. Devolved management requires the co-operation of both parties: central government must be willing to cede some of its authority while the industry - through its organisations - must be willing and able to accept both the opportunity and the burden of added responsibility. Although it may be possible for governments to authorise the devolution of specific functions to responsible fishermen's organisations, the concept of co-management cannot be legislated into being, against the wishes of an unwilling administration or an uninterested fishing industry. Rather surprisingly, not all the user group organisations which might be expected to become involved in co-management appear to welcome the challenge unreservedly, yet, recognition of the value of co-management - of the common interest, shared expectations and reciprocal relationships - must come from within the industry.

The evidence from Denmark, Spain and the UK suggests that the implementation of devolved management may be frustrated by constraints on both sides. For the *industry*, the main constraints would be the fragmentation of interests, the consequent difficulty in reaching consensus over management objectives and appropriate regulatory mechanisms and the uncertain aspiration for a closer involvement in fisheries management. The latter may be no more than an honest appraisal of their limited financial, organisational and professional capabilities. It may also be conditioned by years of decline in the fishing industry and the politics of despair, which makes them reluctant to shoulder part of the responsibility for an ailing industry. Many, but not all, FOs appear locked into a reactive mode of response shaped by the existing system of top-down decision-making. Certainly few FOs demonstrate a sufficiently wide span of interests and expertise to place them at the centre of a comprehensive national or regional fisheries management system.

At the same time, *central governments* (and, more especially, the central bureaucracies) are unprepared to cede responsibility for major management decisions. This should not be confused with the willingness to devolve responsibility for the routine administration of policy implementation to FOs; FOs in an apparently devolved capacity may, in practice, have relatively little influence, sense of responsibility or action potential. Governments argue that the 'disorganisation' of the industry, and doubts about the competence of FOs and their ability to maintain discipline among their membership, would make significant devolution of responsibilities 'unsafe'. Central governments may also take refuge behind the alleged inflexibility of the CFP to argue against change in the domestic policy system. Another particular area of concern, shared by government and industry alike, is for the robustness of enforcement and discipline.

Despite the conceptual elegance of co-management and the need to foster a spirit of co-responsibility, there are a number of risks implicit in the system, together with design challenges. Attempts to reach consensus among a wide range of potentially conflicting interests may prolong the decision-making process and result in weak compromise measures. Co-partnership with the state may serve to disempower and reduce the influence of the industry's professional organisations, where incorporation within the policy process nullifies their traditional 'strategies of resistance' to policy decisions. This may be one reason why some key organisations resist being drawn too deeply into co-management. There is also a significant challenge in obtaining an appropriate balance of influence - user groups cannot be given too much control in key areas of policy making, but too little can devalue the real benefits of co-responsibility in the first instance. Co-management may not lead to a reduction in state regulation nor in bureaucratic procedures: externally imposed regulations may in fact be increased as the state seeks to define, limit and regulate the co-management system itself. Finally, the need for transparency within a co-management system may promote the formal recognition of other agencies (notably those representing marine conservation interests) which broaden the scope of co-management beyond its original remit. Thus FOs may eventually exert less, rather than more, influence on management decisions.

Overall, it would seem that laying down an appropriate institutional framework for co-management may prove to be the easier part of the task. There are more severe challenges, with origins in the very complexity and dynamics of the fishing industry and its institutional and social ecology. Developing co-management as a 'state of mind', built on mutual respect and trust and embracing a common objective, may prove the most difficult. The search for solutions and new opportunities is also complicated by the very diverse conditions that surround the fishing industries within the European Union. Diversity of political cultures, institutional systems and industrial structures at national and sub-national levels compounds the problems of consensual decision-making and the creation of a coherent, integrated approach to fisheries management shared by the central administration and the industry alike. This fundamental diversity can only be handled through a diversity of devolved management designs tailored to particular national or macro-regional contexts.

Notes

1   This Chapter is largely based on the work undertaken by the Universities of Hull, Roskilde and Seville as part of the EU-funded project on Devolved and Regional Management Systems for Fisheries (AIR-CT93-1392, DGXIV SSMA) and co-ordinated by the University of Hull. Part of the Chapter was originally presented as the keynote address at a Seminar on Cooperation in Management of the North Sea and Wadden Sea Fisheries organised by the Dutch Ministry of Agriculture, Nature Management and Fisheries in Groningen, 9-10 January 1997.

2   Currently legislation is being drafted for the extension of Several and Regulating Orders to cover crustacea species (lobsters, crabs).

5 Governing Aquaculture

Dynamics and Diversity in Introducing Salmon Farming in Scotland

JAN WILLEM VAN DER SCHANS

Introduction

This Chapter will discuss the way the introduction of marine salmon farming has been governed in West Scotland and the Shetland Isles. The arrival of salmon farming provided an opportunity for the socio-economic development of these regions, but it has not been unproblematic. These problems are internal to the industry ('souring' of sites due to local pollution, uncontrolled outbreak of pests and diseases) as well as to other uses of the coastal zone (tensions between salmon farming and wild fishing, recreational yachting, visual amenity, nature conservation). From a governing perspective the challenge is to realise the socio-economic benefits while reducing the social and environmental costs related to the development of marine salmon farming. The central question to be addressed is: does the exploitation of marine resources in the coastal zone necessitate a special form of governance? To formulate an answer to this question, we shall discuss the development of marine salmon farming in terms of a number of theoretical perspectives, proposed to help understand problems of governance in relation to marine resource exploitation.

From a *technical* perspective, the development of the coastal zone can be discussed in terms of governing access to and exploitation of common-pool resources (Field, 1987), and also in terms of the 'Tragedy of the Commons' metaphor (Hardin, 1968). From a *socio-political* perspective, the development of the coastal zone can be discussed with the help of a theory of communicative action (Habermas, 1984, 1987), as well as a model of local control over local resource exploitation (Jentoft, 1989; Berkes *et al.*, 1989). This Chapter will first describe the introduction of the salmon farming industry in the UK and its significance for Mainland Scotland and the Shetlands. It will point out the 'common-pool' characteristics of marine salmon farming and provide a more theoretically informed discussion of several governing issues related to the introduction of salmon farming, followed by a general conclusion.

95

Marine Salmon Farming in Mainland Scotland and the Shetlands

Farmed salmon are grown in floating cage structures located in sheltered lochs, bays and estuaries. The fish are fed on a high-protein diet, normally based on fishmeal. Farmers have a variety of chemicals and antibiotics at their disposal to attend to the health of the caged fish. Efforts are made to improve the growth and market characteristics of the fish through programs of selective breeding and genetic engineering. In short, salmon reproduction has been rationalised and has now become a large-scale industrial activity.

In Scotland the first experiments in salmon farming took place in the mid-1960s. Growth was slow in the early years but multinational companies, such as Unilever through its subsidiary Marine Harvest, invested considerable sums trying to overcome problems of nutrition, breeding, disease and equipment design. By the late 1970s, most of the initial technical and husbandry problems had been solved. The prospect of good returns encouraged other companies to invest in the industry and a period of extremely high growth rates began. Production in Scotland increased from 600 tonnes in 1980 to over 40,000 tonnes in 1991 (Scottish Office Agriculture and Fisheries Department (SOAFD), 1991). In 1991, there were some 365 sites in operation, widely dispersed along the coasts of Strathclyde and Highland regions and the western and northern isles (Hebrides, Orkneys and Shetlands). Direct employment in the industry was estimated at 1,014 full-time and 272 part-time jobs (SOAFD 1991). The arrival of salmon farming in isolated communities throughout the highlands and islands has helped reverse a depressing spiral of depopulation and decrease in community services.

Salmon Farming as a Common-Pool Resource

Users of a common body of water, even those operating at large distances from each other, affect - and are affected by - each other's use. Each user is capable of subtracting from the welfare of other users. It is not possible to exclude oneself from the effects of the use of others. The aquatic environment in which fish farms operate can therefore be defined as a 'common-pool resource'. Common-pool resources are resources for which exclusion is difficult and joint use involves subtractability. Governing the development of marine salmon farming can be seen as a particular example of the general issue of managing marine resources and governing the behaviour of its users.

Which natural resource should we take as a starting point for the analysis, and which problems arise with respect to governing the exploitation of the resource? The fish stock itself seems to be the most obvious marine resource to focus attention on. Fish farming is often presented as an attractive alternative to catch fishing, as it is thought to reduce the pressure on wild stocks. From an ecosystem perspective, however, it has been questioned whether salmon farming provides a good example of using the marine environment in an efficient and equitable way. Farmed salmon are fed on fishmeal, some of which is produced from raw fish that may also be fit for direct human consumption. The conversion from fishmeal to farmed salmon is not very efficient, although some progress is made in this respect. There is also concern that salmon that escape from farms could affect the wild fish stocks in negative ways, increasing the competition for food and affecting the genetic integrity of wild salmon stocks.

Another way of looking at the governance issue is to focus on the aquatic environment as such in which fish farms operate. Fish farms affect the aquatic environment in several ways. A variety of chemicals is used in fish farming to control diseases, and some of these substances have 'not been established to be environmentally harmless' (House of Commons Agriculture Committee (HCAC), 1990, Vol. II: 172). Although it seems that public concern over the effects of fish farms on the environment has diminished somewhat as people grow more accustomed to this new industry, it still is quite unclear to what extent the presence of fish farms will affect the marine ecosystem in the long run. Fish farmers do not only affect the environment in general, by releasing toxic chemicals and nutrients, because of their impact on the visual amenity of the coastal region and the like, they also affect each other's performance directly through the spreading of pests and diseases from one site to the other. Because of the 'unnaturally' high density of salmon in cages, the spreading of pests and diseases among wild as well as farmed salmon became a serious problem in Scotland. Fish farmers suffered tremendous losses due to mortalities among farmed salmon because of the uncontrolled spread of diseases. Effective and efficient disease control became one of the most critical factors for the long-term survival of the industry.

Governing Access to and Use of Natural Resources

Theory

Field has proposed a model for understanding the governance of resource exploitation in terms of two general types of governing activities. Governance of common-pool resource exploitation involves collective decisions to regulate the behaviour of those using the resource, and collective decisions to exclude outsiders from using the resource (Field, 1987). These types of decision-making involve 'transaction' and 'exclusion' costs respectively. Field argues that a resource-use governance system is stable when exclusion costs and transaction costs are in balance.

When transaction costs increase relative to exclusion costs, the system will move towards higher exclusivity of use, for example a shift from communal to private tenure. This happens when conflicts between users become more difficult to resolve (higher transaction costs) or when exclusion of non-users becomes more efficient (lower exclusion costs). Fish farming in Scotland has seen a tendency towards higher exclusivity of use at the regional level, as it gradually became more accepted in government circles to acknowledge that intensive salmon farming and conservation of marine ecosystems are two uses of the marine environment which may not easily be combined in one water body. Note that before this change in policy, the official position had always been that salmon farming and nature conservation go together perfectly, obviating need for coastal zones specially designated to marine fish farming development in Scotland. As a result of the policy, there is hardly any loch on the west coast of Scotland that does not harbour a salmon farm. The official recognition that salmon farming and nature conservation could not always be combined resulted in a policy to designate coastal zones as exclusive, either for the development of fish farming or for other uses of the coastal zone such as marine nature reserves from which fish farms are to be excluded.

If exclusion costs increase relative to transaction costs, the system will move towards lower exclusivity, for example a shift from private to communal tenure. This happens, for example, when users among themselves are able to regulate their exploitation more efficiently (lower transaction costs due to innovations in collective decision making and enforcement) or when it becomes more difficult to exclude outsiders who are for example attracted by high profits (higher exclusion costs). In the Scottish salmon industry we can see a tendency towards lower exclusivity of use at the local level. It is not formally possible for established fish farmers to exclude newcomers from applying for a license to set up a fish farm in the same water body, as the decision to accept newcomers is taken by the public authorities, to be discussed below (Crown Estate, River Purification Boards). Established fish farmers can, however, pressure these newcomers to join a local management agreement to co-ordinate actions to prevent the spreading of diseases and decrease pollution, as will also be discussed below.

Field argues that the establishment of a stable resource management regime is not just a matter of economic or administrative efficiency. Eventually it is a matter of political choice. The official authorities may prevent established resource users from excluding newcomers. Official authorities may also have the ability to change collective decision-making procedures in order to facilitate co-ordinated action between resource users. Although Field elaborates on this political dimension, he does not incorporate it explicitly in his framework (Field, 1987: 335-340).

Governing Access to the Coastal Zone

The issue of control over access to the marine environment in Scotland has been quite controversial. Some control over economic activities is normally possible among other things through development plans that are then used as guidance in deciding whether to permit proposed economic developments to go ahead. The establishment of land-based fish farms (hatcheries) falls under normal planning procedures, but local planning control in Scotland does not extend beyond the low-water mark. Hence, local authorities have no statutory control over the number and location of marine fish farms in Scotland. However, anyone wishing to attach farm cages to the seabed must obtain a lease from the Crown Estate Commissioners (CEC). This is a statutory body, appointed by the Sovereign to manage the Crown estates. The Crown estates include some valuable land properties and over half of the foreshore (the area between high and low-water mark) together with the seabed under territorial sea. In the absence of any planning authority with formal control over the marine environment, the Crown Estate Office assumed a voluntary planning role in relation to the development of marine fish farming, although formally speaking it is only the owner and landlord of the seabed. It is this semi-official planning role that has been subject to so much criticism. It was argued that the CEC are neither publicly accountable nor impartial (because they were now both landlord as well as 'planning authority'), and that they lacked the competence required to plan intensive use of what is after all a very complex marine ecosystem.

Until October 1986, leases were issued without any formal consultation. Pressured by the Scottish Office, the CEC agreed to a voluntary consultation procedure to enable interested parties to make their views on applications for fish farm leases known to the Commissioners. Proposals were advertised in the local press and by notices at post offices, and copies of the application were sent to a wide range of interests. In response to criticism on the part of local authorities and other parties consulted by the CEC, in 1988 the Scottish Office undertook a review of the consultation procedures. But the Secretary of State concluded that a convincing case for the extension of local planning control had not been made. He accepted that contentious cases should be more fully discussed and more independently from the CEC. A non-statutory Advisory Committee was established and the body agreed to refer applications to the Committee where one or more of its representative bodies had objected to a proposal during the initial consultation process, and to take full account of the Committee recommendations whether or not to grant the contested lease. The CEC's role in planning marine development remained controversial, however. Despite the fact that several contentious applications for leases had been submitted to the Advisory Committee, it did not meet. The House of Commons Agriculture Committee began an inquiry into the UK fish farming industry in October 1989 and reported in May 1990 (HCAC, 1990, Vols. I and II). In a memorandum submitted to the Agriculture Committee the Convention of Scottish Local Authorities noted that 'the intention of the new consultation procedures was to help resolve the conflicts between the development of fish farming and other interests'. They argued, however, that 'the available evidence suggests that the new procedures are still not satisfactory' (HCAC, 1990: 217). In their report the House of Commons Agriculture Committee rather mildly acknowledged that 'it is only natural for a body taking decisions to draw flak from disappointed parties and we are satisfied, in general, that the CEC has tried to strike a reasonable balance between fish farming and other interests' (HCAC, 1990: iv). They concluded however that 'we are concerned... about the constant controversies surrounding the development of this [aquaculture] industry and by the fact that the CEC has not yet been able to gain the confidence of either environmental bodies or local communities as represented by their local authorities. Although we propose limited reforms at this stage, we do so with the proviso that, if these controversies and antagonisms persist, then it will be necessary to consider a more radical solution to the problem, despite the administrative and even constitutional upheavals this may entail' (HCAC, 1990: xv).

The issue of access control is particularly interesting because the system operating on the Shetland Islands differs substantially from the one operating in Mainland Scotland. Under a private Act of Parliament (Zetland County Council Act, 1974), the Shetland Islands Council (SIC), as harbour authority for the area, has unique powers to regulate all developments in its coastal waters. Anyone wishing to undertake offshore development must obtain a works licence from the SIC and comply with the conditions attached. All applications for works licences are to be advertised and the Council consults widely. Decisions are taken in public and appeals can be made - and have been made in practice - to the Scottish Secretary against the Council's decision, contrary to CEC procedures, which lack an appeal procedure. One way in which the SIC has used its powers to control access to its coastal waters is to try to maximise the benefit of salmon farming to the local economy by only granting works licences to applicants that can demonstrate a 75% local shareholdership. The House of Commons Agriculture Committee concluded that 'the Shetlanders have utilised these powers [of access control] effectively and developed what is very much their own industry' (HCAC, 1990, Vol. I: xiv).

Governing Use of the Coastal Zone

One way to regulate the actions of fish farmers using the marine environment of Scotland is through controlling effluent discharge from their farms. The regulatory framework is provided by the 1989 Water Act. It is, with a few exceptions, an offence to allow the entry of poisonous, noxious and polluting matter, and any solid waste matter into inland, ground, estuarial and coastal waters, unless the entry is authorised by discharge consent issued by the appropriate authority. In mainland Scotland these are the River Purification Boards (RPB), on the islands these are the Islands Councils. When granting discharge consents, the authorities are to ensure that any quality objectives set for the receiving waters are achieved. Assessments have to be made of the effect of the discharge on the receiving waters, taking into account factors such as the composition of the discharge, the quality of the receiving waters and the dilution provided by these waters.

It should be noted that although the formal authority to monitor discharges of fish farms is now clearly established, in practice monitoring caged farms still presents special problems that are different from monitoring land-based sources. Monitoring of floating cages is more difficult because there is not just one point of discharge, e.g. a pipeline, but a diffuse area that needs to be monitored. Another problem is that special equipment is required, e.g. boats to get to the floating cages, which are not needed for monitoring land-based sites. Among conservation bodies there was a concern that RPBs did not have the resources to monitor compliance. In theory, environmental quality standards of certain lochs are translated into carrying capacities of specific sites in terms of numbers of farmed fish. But in practice it is impossible for RPB officials to actually count the number of fish swimming around in a cage. They therefore must rely on company records (administrative control rather than physical inspection). Fish farms beyond a certain size are required to self-monitor their discharges. However, there has been misunderstanding of, and controversy over monitoring methodology proposed and/or implemented by RPBs among fish farmers and inspection officers. It must be noted that to some extent site monitoring is also in the interest of the fish farmer, to prevent 'site souring', i.e. the degradation of the local aquatic milieu due to a compilation of wasted nutrients, salmon excrements, etc. Most fish farmers now hire specialised firms to perform the tests required.

The 'Tragedy of the Commons' Metaphor

Theory

In his famous 'Tragedy of the Commons' Hardin argued that one of the most difficult problems in governing common-pool resources is the lack of criteria for judgements and systems of weighting to realise the 'greatest good for the greatest number'. What is good for one group of users cannot be compared with what is good for another group because the goods are incommensurable. 'Incommensurables cannot be compared' (Hardin, 1968: 1244). Although it may not be possible theoretically speaking, we explicitly or implicitly make the incommensurables commensurate in real life. 'It is when the hidden decisions are made explicit that the argument begins'. Although Hardin recognised the problem of balancing diverse interests, he failed to elaborate on a perspective of how to bring about acceptable compromises. 'The problem for the years ahead is to work out an acceptable theory of weighting' (Hardin, 1968: 1244).

Hardin also argued that if exploitation of a common-pool resource is not regulated, resource users will infinitely expand their exploitative action, and the resource base will inevitably collapse and become of little or no value to anybody - the famous 'Tragedy of the Commons'. 'Freedom in a commons brings ruin to all' (Hardin, 1968: 1244). Referring to herdsmen sharing a pasture open to all, Hardin argued that each individual herdsman will be tempted to increase the size of his herd infinitely as he reaps the full benefits of adding an extra animal while the costs of overgrazing are shared with all other herdsmen. In an unregulated situation herdsmen will surpass the carrying capacity of the pasture; degradation will be the inevitable result. In an unregulated situation the individual herdsman who unilaterally restrains the size of his herd will not reap the benefits of his restraint because other herdsmen will continue to increase the size of their herds to fill in the open space.

There is some, although not complete, analogy between a group of herdsmen using a common pasture and a group of fish farmers using a common body of water. The benefit of increasing stock density in a floating cage is reaped by the individual fish farmer, while the costs in terms of increased risk of outbreak of pests and diseases are shared with fish farmers at nearby sites. These costs include increased fish mortality, and/or increased costs of medical treatment to counter the transmission of diseases through the water body.

To prevent a tragedy of the commons Hardin suggested either to sell off the commons as private property, or to keep them as public property but restrict the right to enter them, for instance through licenses. In the case of the marine environment it is practically impossible to establish ownership of the water body itself, since water is a transient resource. It is possible, however, to claim ownership of the seabed and through this to control access to the water body above the seabed by issuing licenses. In Scotland the CEC own most of the seabed and by virtue of their power to require a license for sea farm moorings they are, at least in theory, in a position to restrict access to the resource and regulate use of the resource in order to balance incommensurable uses of a commons and prevent a tragedy of freedom in the commons.

Governing the Coastal Commons

Interest groups continuously argued that the CEC's role as voluntary planning authority was incompatible with their statutory role as landlords. Under the Crown Estate Act 1961 'it is the general duty of the Commissioners to maintain and enhance the value of the Crown Estate and the return obtained from it, but with due regard of the requirements of good management'. In their own perception of their voluntary planning role the CEC saw it as its task 'to balance the needs of different users, to balance different values', in particular 'to balance conservation and development' (pers. comm.). Conservation bodies noted, however, that CEC guidelines on the siting and design of fish farms in Scotland were 'vigorously promotion-oriented, as should be expected from an enterprising landlord' (SWCL, 1988: 53-54). In their report the House of Commons Agriculture Committee was satisfied, however, that 'commercial profit is not one of the main factors motivating the Crown Estate in judging lease applications' (HCAC, 1990: ix). It can be concluded that, even if the CEC in practice have not taken their financial duty as their only concern, without a clear statutory duty to include other considerations, they remain in a vulnerable position as voluntary planning authority for the marine environment.

With respect to a possible role for the CEC in preventing pests and disease epidemics, the CEC themselves saw their voluntary planning role as pertaining to 'locational' rather than 'functional' (i.e. disease control) aspects of planning (pers. comm.). In practice, however, the prevention of diseases has been an important consideration in CEC siting policies. In their 1987 guidelines on siting the CEC therefore indicated a separation distance between salmon farms of approximately 5 miles to prevent uncontrolled outbreak of pests and diseases (Crown Estate (CEC), 1987: 7). In response to House of Commons Agriculture Committee questions on this issue, the Earl of Mansfield, First Commissioner, noted that 'we are... conscious of the effects of too liberal a granting of leases or licenses which has happened in some places and above all we are very mindful of disease which can literally wipe out a small farmer in a matter of days' (HCAC, 1990: 164). However, despite their awareness of the problem and their general preparedness to contribute to its solution, the CEC have not been able to prevent dramatic disease losses; e.g. Loch Sunart was so heavily contaminated with a combination of sea lice, furunculosis and pancreas disease in the late 1980s that it was no longer regarded a viable location for fish farming. This suggests that either the CEC had more important objectives than preventing diseases, for example maximisation of income from leases, or they did not have the governing capacities needed to prevent the spreading of diseases.

The CEC's standard lease contract with fish farmers contains a stipulation that requires the tenant to 'carry out his operations in accordance with the best and most up to date method of marine farming and to use his best endeavours to keep the stock in good health and disease free at all times'. Such conditions are difficult to enforce in practice. This would require extensive resources, both people as well as equipment, devoted to monitoring and control which were not available to the CEC. It would also be difficult in most cases to prove breach of contract. Apart from these practical difficulties, it has been suggested that the CEC 'have other priority areas' anyway (pers. comm.). In line with their own perception of responsibilities, the CEC placed greater emphasis on enforcement of other lease conditions such as those concerning the exact location of the farm and also the general appearance of the farm. They argue that the Scottish Office 'has disease control responsibilities by statute (...) on land we don't normally deal in detail with husbandry practices of our tenants, so why should we do it at sea?' (pers. comm.).

Although other bodies, such as River Purification Boards, may have more specific statutory duties with respect to disease control, in practice they face the same type of problems as the CEC do in enforcing regulation emanating from these duties. Conservation bodies consistently expressed concern that RPBs and Island Councils were inadequately resourced to deal effectively with this major area of regulation (Scottish Wildlife and Countryside Link (SWCL), 1988: 28; 1990: 13). Even if spot-check visits are conducted, verification can only be through company records. It is just impossible to actually take the fish out of the water and directly verify the quoted numbers. To the extent that regulation of fish farmer action is difficult to enforce once the farm is in operation, it may seem logical to focus attention on controlling the number and location of farms before they establish themselves in the first place. In a discussion paper published in 1990, conservation bodies therefore argued that 'the separation of vocational planning from regulation duties [of the CEC and RPBs respectively] is incompatible with sound resource management. The continuing chaos of powers and duties relating to the use and protection of marine resources must therefore be resolved' (SWCL, 1990: 25).

A Communicative-Action Perspective

Theory

The rationalisation of salmon farming technology is not just a technical process resulting in more efficient salmon production, it is also a socio-political process, as it involves the possibility to create employment and investment opportunities where they did not exist before and the possibility to bring about a more equitable distribution of economic wealth among different regions in the European Union. The rationalisation of salmon farming technology should in any case respond to the needs and wants of people involved, as they articulate these needs and wants in the public and private spheres of society. In Scotland the needs to be fulfilled by salmon farming technology tend not to be expressed in terms of adding farmed salmon to the diet of the local population (almost all of the farmed salmon is exported) but rather in terms of the contribution that the industry can make 'to the social and economic well-being of many remote and vulnerable local communities in the north and west of Scotland' (Scottish Office Environment Department (SOEnD), 1991: 1). But it is a difficult matter to legitimately define the 'social and economic well-being' of a population - the problem of commensurating incommensurables. Habermas would suggest that in modern society the development and legitimisation of solutions to this problem can only be brought about in continuous processes of communication and rational argumentation (Habermas, 1984, 1987).

Habermas defined rationality in terms of the possibility to criticise acts and expressions. Rationalisation therefore refers to the ability to provide valid reasons for acts or expressions. Habermas distinguishes between two types of rationality: goal rationality and communicative rationality. Goal-rational action is about choosing the appropriate means to reach given ends. Communicative action allows for the possibility to question given ends and taken-for-granted means. If it is deemed necessary to establish new ends and develop new means to reach those ends, communicative action aims to realise this through a process of socio-political deliberation, rather than through the application of economic or bureaucratic power. Communicative action is an effort to reach a common understanding of a situation through reasonable argumentation, rather than through manipulation or brute force. Common understanding is brought about if people mutually recognise the validity claims they make in their speech acts.

When parties reach a common understanding of an action situation, this in itself may be enough to bring about co-ordination of action. In modern pluralistic societies however, with a functionally differentiated economic system and a functionally differentiated bureaucratic system, it is unlikely that communicative action alone can do the job of bringing about co-ordinated action in each and every action situation. Habermas therefore accepts that as far as the material reproduction of society is concerned, action may be co-ordinated 'behind the backs of the people involved' through systemic co-ordination mechanisms such as bureaucratic rules and/or the price mechanism. But Habermas maintains that these systemic co-ordination mechanisms must ultimately be rooted in the life-world, i.e. in the final analysis it must be possible to back them up by good reasons. Habermas also maintains that as far as the socio-political and cultural reproduction of society is concerned, this can, and in fact should only be dealt with through social co-ordination mechanisms, such as provided by commonly shared frames of reference, commonly shared norms, and commonly shared value-orientations. Bureaucratic rules and economic calculations cannot be invoked to bring about nor legitimise social action co-ordination. Social co-ordination mechanisms again are legitimate only insofar as they can be ultimately backed up by good reasons.

Habermas' theory of communicative action provides a framework to analyse some of the more complex aspects of governing the exploitation of common-pool resources. Arguments about the best way to manage the resource generally cover the full range of human experience. Interest groups and individuals involved tend to question each other's validity claims with respect to interpretations of facts (e.g. ways to establish maximum carrying capacities of the local resource base), of norms (e.g. rules to guide the granting of leases to different users of the seabed), and even of personal tastes (e.g. ways to evaluate the visual impact of fish farms). In Habermas' view this questioning is rational and legitimate. In their day-to-day communications as well as in their contribution to socio-political deliberations, people bring in their experiences about the objective world as well as the social and the subjective world. The validity claims communicated in their statements do not only refer to the veracity of the facts proposed, but also to the rightness of the acts intended and to the authenticity of the subjective experience manifested in the statements. Habermas' theory of communicative action is aimed at (re-)discovering practical reason in socio-political decision-making. He tries to provide a well-rounded alternative to a one-dimensional conception of reason, i.e. the functional reason as implied by bureaucratic rules and economic calculations, which so often predominates the socio-political reality of modern market-based societies.

Planning in Scotland

Interest groups argued that they had difficulties participating constructively and meaningfully in the CEC consultation process. In a memorandum submitted to the Agriculture Committee the Convention of Scottish Local Authorities noted that 'in sharp contrast to the open procedures and decision making of local planning authorities the Crown Estate Commissioners make their decisions in a regime which is characterised by privacy and a lack of openness'. 'There is no requirement for discussion of issues prior to a decision being taken. There is no opportunity to allow conflicting evidence to be tested' (HCAC, 1990: 215). This criticism can be interpreted as a lack of communicative rationality in the planning process i.e. a lack of ability or willingness on the part of lease applicants and the CEC bureaucracy to give and accept good reasons to substantiate validity claims.

The CEC employ a professional planning adviser, who administers the consultative procedure, and advises the Commissioners whether to grant the lease as applied for, to make modifications, or to refuse the lease completely. The Commissioners themselves make the final decision. The planning adviser weighs the comments and objections from those consulted, and seeks ways in which the original proposal may be modified, or conditions made, such that these objections may be overcome. In cases where several objections cannot be overcome through modification of the lease, the application is refused. Where only one objection cannot be met through some modification to the term of the lease, the importance of that objection is assessed. 'This is a somewhat subjective process taking into account the "significance" or validity of the objection' (Nature Conservancy Council (NCC), 1989: 107). 'The consultation procedure as established does not allow for much dialogue between the CEC and those consulted. Alterations to the lease are often made in order to meet objections, but without checking with the consulted whether the new arrangement would indeed be acceptable. Furthermore, there is no right of appeal against the CEC decision for either developer or consulted' (NCC, 1989: 108).

Habermas understands the economy and the bureaucracy as societal subsystems governed by systemic co-ordination mechanisms. These are spheres of life where actors orient their action towards money and bureaucratic rules rather than commonly shared norms or values. Habermas agrees that communicative action is possible inside the economic or bureaucratic system but he argues that in normal cases people always have the possibility to fall back on some formally arrangement, organised *a priori*, when they fail to reach agreement. Falling back on pre-established formal arrangements would not be much of a problem if these arrangements are ultimately rooted in the normative contexts of the life-world.

The evidence suggests however that the formal arrangements that existed in Scotland to govern the use of the seabed were not capable of providing the legitimacy needed to deal with the arrival of marine fish farming as a new and intensive use of the marine environment. Interest groups consistently argued that the CEC's role as voluntary planning authority was incompatible with their statutory role as landlords. Time and again they pointed out that there was no public accountability and a lack of democratic control over CEC decision making. The Crown Estate Act 1961 provides for a statutory role of the Crown Estate Commissioners as landlord with an obligation 'to maintain and enhance the value of the Crown Estate and the return obtained from it'. It did not provide for a statutory role for the CEC that included other dimensions of stewardship as embodied in pre-capitalist feudal law, such as duty to the community. The Crown Estate Act 1961 only formalised part of the Commissioners' duties as landlord, namely their duty to realise an economic revenue, leaving the possibility of other more comprehensive duties of the Commissioners unattended. Following Habermas, one cannot object to a formalisation and commercialisation of the role of the landlord as such, but this should not infringe upon the capacity of people to participate meaningfully in the socio-political deliberation leading up to regulation controlling the exploitation of their natural environment. On land, a framework of formal rules had been developed to guide and, if necessary, counterbalance the economic pressures to rationalise land use; that is, planning of land use was brought under democratic control. In Scotland no such framework has developed for planning the use of the marine environment. In a democratic society it is not an appointed administrative body, the CEC, but an elected political body, Government, that is ultimately responsible.

Planning in the Shetlands

By virtue of its role as harbour authority of the waters around the isles, the Shetland Islands Council (SIC) developed a planning system that differed in some interesting ways from the system prevailing in West Scotland. Anyone wishing to undertake offshore developments must obtain a license from the SIC. All applications for works licences must be advertised. The Shetland Islands Council consults directly with all the relevant interest groups in conservation, fisheries and the relevant public departments. The Community Council in the area is informed. The Shetland Salmon Farmers' Association (SSFA) is informed, but does not comment on individual applications, rather on aspects of general policy. Decisions on applications are taken by the Council with all observations or objections being considered in public. Both applicant and objectors who are aggrieved by the Council's decision to refuse or grant a works licence have a right of appeal to the Secretary of State for Scotland. A local inquiry is then held and the Secretary of State publishes a decision after considering the inspector's report. In this way the Secretary of State can overturn SIC's decisions, and has indeed done so on several occasions.

Apart from this formal consultation process, the SIC encourages informal 'pre-consultation' between applicant and interested parties to discuss and work out difficulties before the application reaches the Council table (HCAC, 1990: 268). As all applications pass through this 'filtration system' they are less likely to conflict with other uses of the marine environment.

The key difference between the SIC's and CEC's consultative processes is that the Shetlands have direct and open communication between all interested parties. There is an opportunity to test each other's validity claims and to work out compromises. The final decision is made in public and both the applicant and councillors table a complete package of observations and objections. Broadly speaking, the works licence procedure in the Shetlands provides for public participation similar to regular planning procedure, but different from CEC consultative procedure.

The Shetland Islands Council is not only concerned about providing arguments to validate decisions on individual cases but also to validate the Works Licence Policy in general. In 1985 the Council appointed a Fish Farming Working Group which provided a forum for exchange of views between the Council and the industry. Representatives of the Nature Conservancy Council (now Scottish Natural Heritage) and the Highlands and Islands Development Board (now Shetland Enterprise) were also invited to attend meetings of the Group.

In consultation with the Fish Farming Working Group the SIC developed a Works Licence Policy, aiming to encourage the development of a salmon farming industry in Shetland, to make salmon farming a key element of Rural Development Strategy and to maximise the benefits to the local economy. Note that it was not until 1982 that the first salmon farm came to Shetland, hence the Shetland Islands Council could learn from experience gained by others. 'Quite a lot of policy considerations were evolved through inspection and examination of what was going on elsewhere particularly in Norway and also in Scotland at that time' (HCAC, 1990: 272).

Co-ordinated Action in Disease Prevention

Above we have discussed the need for co-ordinated action to govern the development of the salmon farming industry in relation to other users and uses of the marine ecosystem. Now we shall turn to the need for action co-ordination of salmon farmers sharing a common water body to prevent the outbreak of pests and disease epidemics. This is an interesting issue because it provides an example that even to solve a rather technical problem among salmon farmers, it is almost impossible to rely on bureaucratic rules or economic incentives alone to bring about co-ordinated action. The action situation is so complex and dynamic that co-ordinated action can only be realised and sustained (if at all) if participating actors share a common understanding of the action situation and/or are prepared to engage in communicative action to work out differences of interpretation that either exist or may emerge as they go along.

As noted earlier, there is a very realistic possibility of a Tragedy of the Commons if pests and diseases are transmitted through the water body from one farm site to the other. The problem is further complicated since resistance to treatments can be transmitted throughout the water system. To prevent epidemics, fish farmers make use of various chemicals and antibiotics. They have also developed preventive strategies such as stocking at lower densities with smolts of guaranteed health status, site rotation and site fallowing to break the cycle of contamination. However, as most loch systems in Scotland contain multiple farming operations owned by different companies, prevention of epidemics is only possible if these companies co-ordinate their activities at the different sites. Little is gained if a single farmer reduces stocking density, uses smolts of guaranteed health status or leaves his site fallow, while other farmers continue to grow contaminated high density stocks at contaminated sites nearby. Much is gained, however, if neighbouring fish farmers agree to co-ordinate stocking densities, disease treatment and fallowing intervals.

Despite this 'overriding logic', in 1992 there were still relatively few examples of agreements between farmers operating at neighbouring sites to prevent the spreading of diseases (*Fish Farmer*, 1992: 3). This has been explained from the fact that marine salmon farming is still a relatively new phenomenon. As one site manager put it: 'We have been so busy with running the farms, we haven't really looked at how to organise the business'. The lack of enthusiasm to co-operate has also been explained from historic factors. Companies that invested substantially in research and development to improve husbandry techniques, tend to be very cautious to share this experience with neighbouring farmers, whom they perceive as competitors. Given extensive press coverage in the past, companies are none too eager to admit that they have disease and/or parasite problems, as they do not want to harm their public image. Fish farmers also tend to believe that information on quantity and time of harvesting can be of commercial value to neighbours as competitors. Furthermore, smaller companies, operating at only one or few sites, may not be able or willing to fit their production cycle to the requirements of bigger companies sharing the same loch. Lastly, there are sceptics who just do not believe that good intentions laid down in an agreement will hold up in the face of any really serious problems. 'You may share information and all that, but after all if your fish is going to die you try everything whatever agreement you have. You try every treatment available and you don't care about your neighbour building up resistance' (pers. comm.).

However, as the benefits of action co-ordination in disease prevention are so obvious, an increasing number of companies operating nearby sites have been able to work out an arrangement all can agree to. In 1991 some 84 out of 286 sea-cage sites reporting to the Scottish Office Agriculture and Fishery Department's industry survey were involved in some kind of joint arrangement (*Fish Farmer*, 1992: 3). It is interesting to note that although these joint 'area health agreements' are increasingly formalised it is practically impossible to fit them into a system of monetary incentives and/or bureaucratic controls. This is an example of a transaction where performance evaluation is so difficult that neither monetary incentives nor bureaucratic controls would be able to prevent opportunistic behaviour, yet task interdependence is so great that co-ordinated actions must be brought about. In such a situation, only co-ordination of action based on a common understanding can do the job.

Area health agreements can be successful only if there is some minimal level of trust. 'There was no point in going on a witch-hunt, as has been done in the past... to attribute blame to one company or another for introducing furunculosis to the area. We have to face it, [the disease agent] can be transmitted through the sea, whether directly or indirectly, easily and over large distances' (*Fish Farmer*, 1992: 10). As salmon farming takes place in an open water system, it is difficult to prove that the outbreak of a pest or disease is due to the fact that one of the parties to the agreement failed to meet his obligations. It is practically impossible for parties involved to monitor each other's compliance to the agreement. This would involve checking for example whether your neighbour's new smolts are indeed of the specified health status, whether he is indeed only using the specified treatments, not a 'cocktail'. The possibility to take legal action and get compensation for damages incurred is seriously undermined anyway by the fact that the other party can always argue that the disease came from wild salmon or from any other source beyond his control.

Therefore, 'voluntary participation and a co-operative attitude are and will remain a very critical factor in the success of these agreements' (pers. comm.). Prevention of epidemics will only work if companies operating neighbouring sites have a common understanding of the nature of the problem (e.g. obtain advise from veterinary professionals of similar educational backgrounds), share a common motivation for co-ordinating their action (e.g. share previous dismal experiences), and are committed to helping each other in finding practical solutions (e.g. provide an alternative site to the small company to allow him to come into phase on stocking). Furthermore, parties must be willing and able to work out mutually acceptable solutions to unexpected problems that will inevitably surface. Hence, parties must approach each other in a communicative mode.

Although public authorities can play and indeed are playing a facilitating role in area health agreements, it is difficult for them to actually use their formal authority, even if they had one, to force parties into an agreement. This is attributable to the very same practical reasons: fish farmers themselves have problems to monitor compliance to the agreement and the companies involved will only accept extra duties imposed upon them by public bodies if they are also given extra rights - the legitimacy problem. This would involve, for example, the designation of formal arbitration authority to an independent body to settle disputes between participants in an agreement, or the establishment of a publicly funded compensation scheme for mandatory slaughter of afflicted fish. If participation in an agreement with established fish farmers is formally required as a pre-condition for a lease or discharge consent, prospective fish farmers may enter such an agreement only to obtain the lease or consent - once they are in business they may be indifferent to making the agreement a success (pers. comm.). Thus, it is clear that area health agreements will not work if the actions of participants are inspired by short-term, opportunistic considerations rather than by authentic willingness to co-operate - in Habermas' terms if they are inspired by 'strategic action' rather than 'communicative action'.

Local Control over Local Resources

Theory

Hardin has suggested either to institutionalise private property or to allow a dominant role for the state to prevent resource collapse. But a wealth of empirical evidence shows that resource users have been able to restrain use and exclude outsiders from entering the resource base to prevent overexploitation, and that they have been able to do so without relying to any large extent on private property rights or state intervention (Feeny *et al.*, 1996, Berkes *et al.*, 1989). On this account, the Tragedy of the Commons results from the absence or collapse of local control over local resources. Community-based management may, it is argued, be better placed than state management or market co-ordination to bring about co-ordinated action and prevent the collapse of the local resource base.

Local users often have better knowledge of the local ecosystem than distant state bureaucrats, and/or distant Boards of Directors of resource exploitation companies (Jentoft, 1989). Local users, born and raised in the local community, may also share a common worldview, common norms and a common value orientation vis-à-vis the local resource base. Thus community membership provides for informal but more comprehensive ways of co-ordinating exploitative action. Local users may also have a different, more long term, perspective on managing the local resource base. They may value it as a 'collective life insurance', a means of long-term communal survival rather than a private investment project, a means of maximising short-term individual revenues.

Arguments in favour of community control over local resources often assume that local communities are almost completely dependent on the sustainable exploitation of their local resources. In such situation the local community can be expected to do everything it can to manage the local resource base sustainably in order not to destroy the very basis of its long-term survival. But in a modern society context, integration in the broader socio-economic system has made local communities less dependent on sustainable exploitation of their local resources for their long-term survival. Hence, there are fewer built-in incentives to exploit local resources in a sustainable way. Also, local users may have insufficient knowledge of the exact carrying capacity of their local ecosystem anyhow, or they may be tempted to overestimate this carrying capacity, especially when the economic rewards of exploiting that resource base are very high. Local users may also have little consideration for the ecological or esthetical value of their local ecosystem within the broader framework of the national or global ecosystem. Furthermore, in the context of a pluralistic society resource users can be expected to differ more widely than community members in a subsistence economy context, in terms of their worldviews, norms and value orientations vis-à-vis the local resource base. It cannot be taken for granted, therefore, that the self-governing capacity that may have existed in traditional resource dependent communities, still exists in a modern-society context. This is why proponents for more local control advocate a distribution of management responsibilities between local communities and the state where delegation of management authority to the local level is the rule and retaining management authority at the higher level is the exception (McCay and Jentoft, 1996).

Habermas' theory of communicative action allows us to understand the issue of 'local control' in terms of the opportunity of local people to participate meaningfully in the socio-political deliberations which result in regulation concerning the exploitation of the local resource base. Formulated in this way, it is not necessary that local people share common worldviews, norms or value orientations about the local resource base, nor is it necessary that local people and non-local parties are in agreement from the outset. What is required is a willingness (and ability) of all parties involved, both non-governmental and governmental, to be engaged in communicative action, to give and take good reasons for claims made and positions taken in the negotiation and decision making process. From a communicative-action perspective, co-ordinated action at the local level doesn't necessarily depend solely or mainly on social co-ordination mechanisms, as has been suggested by community-based management advocates. Bureaucratic rules and monetary incentives may be needed anyway to bring about action co-ordination. What is required from a communicative action perspective, however, is that bureaucratic rules and monetary incentives can and will, in the final analysis, be backed up by good reasons.

Local Control in Scotland and the Shetlands

The rapid expansion of the salmon farming industry in West Scotland, and the predominance of multinational companies in operating the industry, has been described in terms of a 'colonisation of the coastal frontier' (pers. comm.). In the words of the conservation bodies 'many communities feel hard done-by in the rapid uptake of available sites by large outside concerns which have the resources to keep well informed and to place a large number of exploratory applications' (SWCL, 1988: 47). Yet it cannot be concluded that societal groups in Scotland were unable to exercise any control whatsoever over the development of the salmon farming industry. People may have experienced the arrival of the industry as an invasion, but they did not become totally inactive as a consequence. There was extensive media coverage, effectively bringing many of the controversies to public attention. Voluntary bodies concerned with wildlife and countryside conservation in Scotland formed an association in February 1987, Scottish Wildlife and Countryside Link, to become more efficiently organised to counterbalance the rapid expansion of the industry. Similarly, the local authorities in Scotland that were affected by the marine salmon farming industry joined forces to evaluate and re-design development control arrangements in the Convention of Scottish Local Authorities. Although conservation societies, local authorities and other interest groups may not have been immediately effective in changing the planning system, they were quite successful in bringing forward their arguments, and informing the general public.

The Shetland Islands Council used its special powers to institute a restricted access policy, to create an industry that was locally owned and integrated in the local economy as much as possible. Local people often own salmon farms in partnership. A relatively large portion of people are only employed in salmon farming on a part-time basis and also work in the islands' other industries such as catch fishing. In a memorandum submitted to the House of Commons Agriculture Committee, the Shetland Islands Council noted that the statutory framework of the Zetland County Council Act 1974 'has involved the Council in direct control of the local salmon farming industry, from its inception' (HCAC, 1990: 266). The Council concluded that 'use of those powers to regulate fish farming has proved very satisfactory and could provide a useful model for emulation, if any change were contemplated in the regulatory powers which apply elsewhere in the U.K.' (HCAC, 1990: 265).

Habermas' theory of communicative action allows us to describe the Shetlands as a more 'integrated' society than the Scottish Mainland, without having to conceptualise it as a more 'traditional' community. From the interviews conducted, there is no overriding evidence that the people in Shetland have a different, more homogeneous, more culturally embedded perspective on economic action than people in other parts of Scotland. The Shetlands may be 'peripheral', but not 'isolated'. Broadly speaking, Shetland salmon farmers have access to the same technical, financial, and legal expertise, and the same if not even better salmon farming technology. They compete on the same world markets and share the same political system but for the special powers of the SIC.

People interviewed in Shetland, it must be admitted, did refer to their 'islands tradition' to explain the rather unproblematic societal integration of salmon farming. But such a general reference to culture can hardly account for the resolution of conflicting interests in particular situations, nor can it serve to explain the opportunity for institutional choice for the Shetland Island Council that did not exist for local authorities elsewhere in Scotland.

An official of the Shetland Salmon Farming Association may refer to the Shetland salmon farmers as sharing a strong sense of 'island community'. 'You don't stand out too much. You may need your neighbour tomorrow.' (pers. comm.). But it must be borne in mind that there are very good reasons for an interest representation group to close its ranks and project an image of consensus among its membership: Talking to a Shetland salmon farmer in private, you may find that it is apparently part of the same 'islands mentality' to be individualist, striving for the least possible dependency on your neighbour. Apart from that, there turn out to be as many issues of contention within the local community as there may be among people from different regions of the UK who are attracted to the high returns of salmon farms in West Scotland. It is not the absence of conflicting interpretations or interests but rather the way these conflicts are dealt with that constitutes the basis of legitimate decision-making. The system of formal rules operating in Shetland gave people the opportunity to argumentatively construct and re-construct their understanding of the action situation with respect to governing the development of the salmon farming industry locally.

Conclusion

This Chapter has discussed several ways of approaching governing situations in salmon farming in Scotland and the Shetlands. The Chapter demonstrated that different theoretical perspectives can be deployed to analyse and understand important governing issues in relation to marine resource management. The examples should not be read as testing hypotheses or addressing specific questions raised by these theoretical perspectives. More thorough analyses have been presented elsewhere (Van der Schans, 1993, 1996, 1997), and will also be part of a forthcoming study (Van der Schans, forthcoming). This Chapter has argued that the introduction of marine salmon farming cannot be looked upon as an isolated activity to be dealt with by a single regulatory agency. Rather, it requires co-ordinated action of public authorities and private parties to cope with a diversity of interests in relation to a complex and highly dynamic marine ecosystem. Theoretical perspectives help to highlight a number of technical as well as social-political governing issues, illustrated by experiences from two regions with different ecological as well as institutional settings. Theoretical reflection not only lends more specific insights on factors that may explain the different outcomes in different contexts, it may also provide a basis for applying the insights gained to decision making in practice, in order to facilitate the difficult choices that must be made to govern the sustainable exploitation of the coastal zone.

Note

1   The research on which this Chapter is based is also part of a forthcoming Dissertation. It was made possible by a grant of the European Commission, Directorate General for Fisheries, Directorate of 'Structures', Fisheries and Aquaculture Programme. For more details on the research results, the reader is referred to the Final Report Contract FAR MA 3.757, DG-XIV, European Union.

6 Adding Quality to the Fish Chain

How Institutions Matter

JAN WILLEM VAN DER SCHANS, KAIJA I. METUZALS, NYNKE
VENEMA AND CARLOS IGLESIAS MALVIDO

Introduction

Previous Chapters have argued that managing the harvesting side of the
fisheries chain requires institutional arrangements that reflect the specific
conditions prevailing in the type of fishery at hand and the socio-cultural
settings into which it is introduced. The diversity of conditions across
fisheries and societies on the European scene is far too great to expect a
single model to be generally applicable in all settings. Neither is it realistic to
assume that centralised governance is suited to address all this diversity. A
standardised solution to complex problems and diverse situations may work
in random cases, but in most instances it should not come as a surprise when
hierarchical solutions fail.

A solution advanced throughout this book is co-governance, i.e.
institutions that allow users to effectively participate in decision-making by
applying their experience-based knowledge and sensitivities to challenges as
they are felt in the daily lives of their communities. Such institutions also
imply devolution of management functions to user organisations at the meso-
level, such as the community or regional levels. Furthermore, co-
management provides incentives for co-operation among users and co-
operation between users on the one hand and government agencies and
research communities on the other. Thus, co-governance relies on shared
responsibility for management functions and interactive decision-making
with respect to collective goods, such as healthy fish stocks.

If this argument is valid for the harvesting side of the chain, it
should also be relevant to the chain as a whole. First of all, co-ordination
problems and, hence, management demands, emerge not only along the
horizontal axis, i.e. among actors at the same level in the fisheries chain,

such as fishermen: they also frequently take place along the vertical axis, i.e. among transaction partners at various levels in the distribution chain - between fishermen, fish-processors, and traders of fish. Market saturation may affect chain participants in the same detrimental way as the exhaustion of fish stocks may. Notably, the two 'tragedies' occur for similar reasons. Sustaining the resource base is a collective-action problem, just like sustaining the quality of the fish product is. In both instances, action co-ordination in one way or another is essential. Ruined fish stocks, market flooding, and inferior fish quality may result from individual rational actions. The Prisoners' Dilemma applies in all cases, which suggests that the problem could have been avoided if micro-decisions were co-ordinated more effectively in the collective interest. Here, co-operation and interactive co-governance among the chain partners is an alternative to the hierarchical command-and-control intervention by external authorities, such as the state.

Secondly, prevailing institutions established for the purpose of vertical co-ordination of transactions throughout the chain, such as Producers' Organisations, are at least as diverse as those among harvesters (see also Chapter 4). Despite the fact that they find a common basis in European law, the structure as well as the practice of POs differs among countries and fisheries. If it is hard to track down the complexities of social and economic relations between fishermen, it emerges from our research experience that it is even more complicated to discern the patterns of exchange between catch and market. They differ not only from country to country according to historical, cultural and political developments, but also from fishery to fishery and from fish product to fish product.

Again, forcing a standardised model upon the fisheries system regardless of context may accidentally produce positive results. From a governance perspective, homogeneity is often preferable to diversity, if only for ease of control. Whether homogeneity promotes better adjustment to market demand is, however, a different story. The general expectation is that structural diversity in the organisation of chain relations is a result of market diversity. In other words, organisational set-ups for the co-ordination of exchange are made to fit the dynamics that occur within market niches. Both the variety of market niches and the particular dynamics that characterise them, require distributional flexibility on the part of chain participants. Standardised hierarchical structures are normally less responsive to fluctuations within and among market niches than institutions that are heterogeneous, autonomous, and embedded in the socio-cultural context circumscribing the actions of the transaction partners.

This Chapter has two objectives. First, we set out to show some of the diversity in chain organisation in European fisheries. Rather than attempting to reduce this diversity in order to make it fit a more homogeneous, preconceived scheme, we argue that governance should treat this diversity more pragmatically, take it as a premise and view it as an asset to be learned from and built upon. This requires co-ordination to be handled in a decentralised, yet orderly manner, through networks and strategic alliances rather than hierarchies. The co-governance model has much to offer here.

Furthermore, we show how a particular governance problem, the issue of quality control, can be handled co-operatively by interacting chain participants. The quality issue is a prime example of problems that must be addressed individually as well as collectively, i.e. by each participant in the chain as well as by all participants interactively. There is a temptation to free-ride but because all stand to gain something, there is also an incentive to co-operate. With respect to quality, the chain participants are particularly vulnerable. They are directly dependent on each other as quality lost early in the distribution process can not be recovered at a later stage. Although the government is involved in guaranteeing minimum quality standards, in most cases it is left to the chain partners themselves, whether they co-operate on higher-level quality goals or just comply with the minimum standards. Interestingly, in some European countries Producers' Organisations have taken on a more pro-active role in quality enhancement throughout the chain. The sections below will discuss some of the lessons learned.

We shall start out by elaborating on the theoretical position adopted for this Chapter, then proceed to explore the diversity issue by presenting four case studies of catch and market systems from various settings in Europe. Quality promotion and control from a governance perspective are discussed next. In particular, we wish to address the issue of how quality standards could be negotiated to fit the great diversity of chain relations that exist among fisheries systems. In other words, how can all this diversity be accounted for when promoting quality standards on a Europe-wide scale?

Theoretical Perspective

Chain relations could fruitfully be analysed from an institutional perspective as developed within organisational theory, sociology and economics. This perspective argues that institutions are key in determining the outcome of social and economic processes. Institutions provide for more or less fixed sets of routines, roles and rules. Thus, institutions constrain as well as enable social action due to the fact that they decrease uncertainty and create consistent expectations between interacting individuals, such as people involved in economic exchange (Scott, 1995; Powell and DiMaggio, 1991). In this respect, institutions fill the gap between the social system on the one hand and adaptations at the micro level on the other. They bring stability and predictability. The costs involved in arranging transactions may be cut if institutions are properly designed. In this way, institutions have social as well as economic value.

From this perspective it is argued that what is considered rational behaviour is contextually defined and institutionally anchored (Selznick, 1992). Egocentricity and opportunism in market transactions (Williamson, 1975) may seem rational from a strictly individual point of view, but it may be condemned among trading partners on moral grounds, even within markets where loyalty is assumed to be low. As Kenneth Boulding points out (1968), market behaviour is guided by moral norms; markets could not function unless there is some basic agreement on ethical standards.

Within this perspective, it is assumed that social institutions governing economic actions are subject to change and can be influenced by social interaction. Institutions are never determined for once and for all, despite the fact that they are usually taken as given in day-to-day interaction. In reality institutional change happens all the time, even if marginally and slowly. Sometimes, however, institutions are subject to purposeful intervention and design, as we shall see below.

Institutional economic theory assumes that the cost of providing the institution itself is an important cost to be taken into account. To work effectively the parties involved must also accept institutions as legitimate. The institutional argument is that if transaction costs are low, the actors involved will be able to work out acceptable solutions to their co-ordination problems among themselves. Institutions should therefore facilitate this capacity for self-management. As transaction costs increase, internalisation of transactions within firms through vertical integration are more likely to be cost-effective. However, there are also costs to vertical integration as firms get larger and less adaptive to the local circumstances. The larger the firm, the more standardised and formalised the administrative structure gets. Therefore, in many situations coalitions and networks of firms are preferable. In other words, instead of relying on spontaneous co-ordination, as in the marketplace, or co-ordination through command and control principles as in the hierarchy, network systems rely on co-operation, mutual adjustment and communication as co-ordinating mechanisms. We believe the latter co-ordinating vehicles are particularly relevant to fisheries settings.

Governing Diversity in the Fish Chain

This section will describe the role that Producers' Organisations play in the governance of the fish chain in various contexts. POs are based on European law, and their main purpose has been to administer the price support system. They draw their membership from among fishermen groups, organised around fisheries sectors, fleets, ports, regions etc. In addition to the role they have according to EU regulations, they have in many cases adopted additional roles and functions. These will be described in the case studies summarised below[1]. The case studies include Vigo in Galicia, Spain, Peterhead in Scotland, the Shetlands and various fishing communities in the Netherlands.

Vigo, Northwest Spain

The Spanish case study reported below demonstrates the great organisational diversity that may exist at the catch level. The fishing industry is organised as a number of associations and Producers' Organisations, each representing fishermen by gear type, fishing area and fish targeted. These organisations play an important role in catch co-ordination and more recently in improving the marketing conditions by way of planning the catching process as well. Thus, Producers' Associations and Organisations are expanding their responsibilities beyond their original task assignments. The Vigo case also suggests the importance of building on existing Organisations rather than establishing entirely new Organisations in devolved systems of fisheries governance.

Vigo is one of the largest fishing ports in the world. Situated in the province of Galicia in northwestern Spain, its estimated population is 274,574 (according to the 1991 census). The surrounding conglomerations included, it supports approximately 431,450 people. However, from the economic point of view, Vigo shares many disadvantages with Galicia, one of the poorest regions in the EU. The Gross Domestic Product (GDP) of Galicia is 63% below European average. There is an automobile industry as well: Citroën, the car multinational, has a plant in Vigo that provides direct employment to 7000 people as well as 1400 people in related local jobs. However, shipping is the main activity of Vigo, both passenger, cargo and fishing, fresh and frozen fish, molluscs and cephalopods.

Vigo has a modern port facility that can handle a quantity of frozen fish, an estimated three times the volume of fresh-fish landings. The capacity of the cold storage plants installed in the port is among the finest in the world (Vigo Port Authority Board, 1994). There are an estimated 50,000 people working in activities related to the fishing industry such as the shipyard and sea traffic operation.

In 1994, fresh fish auctioned in Vigo totalled 91,411 mt. For the past 50 years, 60-80,000 mt of fresh fish have been sold annually at the traditional El Berbes harbour designed for fresh-fish landings. Fish dealers and traders buyers organised a network not only catering to the whole of Spain but also to Italy, Portugal, and France. The local canning factories are also supplied with fresh products from the market of Vigo. Frozen fish is handled by either freezer ships, most of which are based in Vigo, landing 61,029 mt in 1994, or by reefer cargo vessels that transport the frozen fish from the fishing grounds to the port. About 252,525 mt were transported in this way. Total fish handled in Vigo (fresh and frozen) was 404,965 mt in 1994; 12.58% more than for any other European port.

Vigo also seats the headquarters of Pescanova, one of the world's largest fishing companies, a leader in processed food. There are also very large and powerful Producers' Organisations. One of the main associations in Vigo is the ARmadores de Pesca del Puerto de VIgo (ARVI) co-operative, which was formally established in 1964 although it did not start its activities until 1973. Mr. Llanos Suarez, the Assistant manager for ARVI explained how the PO came into being: 'We were forced by the EU. The *cofradia* was not accepted by the EU'. There are now nine fishing vessel-owner associations: ANAMER, ANAVAR, ANAPA, AGARBA, ARPOSOL,

ARPOSUR, ARPOAN, ARTEMAR and CERCO, as well as two Producers' Organisations (POs) in this organisation: OPPC3 and 4. This association in Vigo comprises at the moment 428 ships, 250 of which are freezer trawlers, 6 cod trawlers, 172 fresh fish trawlers of a total of 186,677 GRT and 8,536 crew members on board (ARVI, unpub. doc).

Their main functions are unloading fresh or frozen fish, cleaning shipholds, and storing ice after unloading. There are also the services of hull inspection, removing propellers, cleaning sonar detectors and purchasing supplies, from electronic equipment to nets and hooks.

As well as these practical aspects, there are the regulations to keep up-to-date. Fishing is a complicated matter: the rules and regulations change and keep changing. The PO has a specialised service to look after just that, from the different fishing grounds where the distant water fleet might go. All catches and fishing efforts of the fleet in the different countries where it operates are compiled. Some of these areas are the NAFO (North Atlantic Fisheries Organisation) areas, off Namibia, South Africa, Morocco, Portugal, the Spanish coast etc.

This PO is said to be the strongest and most powerful representative of the Spanish Fishing Sector and even of the European Union (ARVI, unpub. doc.). The 428 associated ships (ranging from a 9.6m minimum to 103.4m maximum length with an average length of 41m) are associated with 292 associated corporations or firms.

Other functions of the organisation are to keep all the members informed of events in the fishing sector that may affect them. The organisation provides the producers and owners with a regular publication, newspapers and specialised magazines, it has a press office and links to the media. One service lends its members advice on the subject, another department advises on labour issues, trade unions or social-security matters.

The nine sub-organisations in ARVI are grouped by gear and/or fishing area. ANAMER, short for Asociación Nacional de Armadores de Buques Congeladores de Pesca de Merluza (*merluza* is Spanish for hake), is an association that covers the whole country and includes all Spanish refrigerated ships dedicated specifically to fishing hake and cephalopods (squid such as *Illex* and *Loligo*). Their main fishing grounds are located in South and Southwest Africa, Namibia, Southwest Atlantic Ocean and the Falkland Islands. Hake is processed and frozen on board, either whole, gutted and headless, or in fillets, skin on and skinless, either in fish blocks.

There are also by-products such as roe. Squid (either *Illex* or *Loligo*) is frozen whole or in the 'eviscerated tube' form. This association represents the most technologically advanced boats in the Spanish fleet. The total length of these vessel ranges from 35.45 to 90.3m; their average voyage takes five months.

ANAVAR, another nation-wide association, has members in the freezer trawlers that alternate their activity in different fishing grounds and they also catch different species. The vessels are of medium tonnage and traditionally frequent the North Atlantic Ocean as well as grounds off the coast of Namibia. The main species targeted are squid, octopus, cuttlefish, hake, flounder, witch flounder, halibut, cod and the common prawn. The trips that these vessels make are on average three to four months in duration.

ANAPA is the national association for longliner freezer vessels, although they can also work with other fishing gear. They target swordfish, mackerel shark and other sharks. Their fishing grounds are in the Third World: off Senegal, Guinea and Gambia and elsewhere. These vessels make 45 to 90-day trips as they follow their migratory species.

AGARBA is the association for bacalao or cod. There are six ships in this category. Their average size is 47m, and on average 23 crew can be carried on board. These vessels fish in NAFO waters as well as other international waters. The cod is processed, salted and dried.

ARPOSOL is an association of vessels that fish in EU waters, mainly near the coast of Ireland, United Kingdom and off France. The main part are trawlers but there are also longliners. The average trip is between 10 and 18 days. These species are megrim, monkfish, octopus, hake, ray, haddock, squid, flounder, horse mackerel, mackerel, ling, saithe, Norwegian lobster, pollack, red gurnard, seabream, blue whiting and more. The majority of the companies associated with ARPOSOL have a family character, as revealed by the motto: one company, one vessel. These are also the vessels which fish in the Grand Sole Bank as seen from their name: Asociación Provincial de Armadores de Buques de Pesca de Gran Sol de Pontevedra. There are 71 ships in this association with an average length of 30m. These ships can carry on average 16 crewmembers. The ships here and in AGARBA are also older than the rest of the fleet. Ships in both organisations are on average 20 years and over.

ARPOSUR is an association of eleven trawlers which fish off the Spanish and Portuguese coasts. The main thrust is for the fresh-fish market, including monk fish, small hake, octopus, pout, horse mackerel and

Norwegian lobster. The average trip varies from one to four days in Spanish territorial waters and up to ten days for trips in coastal waters of Portugal. These are older ships, on average 22 years, with an average length of 30m.

ARPOAN is the association of longliners. Their main catch is the short-fin mako shark, swordfish and tuna. Vessels in this category sail long distances, chasing their prey. As written in the documentation: 'We cannot confirm that the vessels of this association have concrete (sic) fishing grounds, even when their activity is carried out mainly in the oceanic waters adjacent to the Iberian Peninsula, with sporadic trips to Spanish territorial waters and even reaching the North African and EU waters.' There are 28 member ships, averaging 27m in length.

ARTEMAR has gill-netting vessels. The main species fished for are monkfish, skate, crawfish and sole. The traditional fishing grounds are the Galician littoral zone and North Africa. The trips that these vessels make can vary between two and twenty days. There are six vessels of an average length of 21m in this category. The number of boats and associated firms in this category are the same: six and six.

CERCO is the last association, grouping the purse seiners. They are vessels of different sized ranging from 5 to 40 GRT, and upwards from 13m in length. The fleet is mainly concentrated in Spanish territorial waters, especially in the Galicia Ria. The main species is sardine, a pelagic fish. But mackerel, horse mackerel and sprat are also caught. The trips are day trips lasting from 4 to 12 hours. There are ten ships in this association and ten associated firms.

OPPC3, a PO that draws its members from the whole country, was established three months after the effective application of the Common Fisheries Policy in Spain. It represents 60% of the total freezer fleet. The main species it goes after are hake and squid. These three species represent 80% of total catch. The main objective of the organisation is to 'rationalise fishing' and to obtain better sales conditions for their production. The PO has an annual catch plan with the capability of adapting 'the volume of the offer to the necessities of the market at each moment'.

OPPC4, a PO established in 1986, serves the fresh-fish market in Ponteverda. The main species are hake, megrim, monkfish, pompano, swordfish, sardines and mackerel. The vessels fish in EU waters, Atlantic Ocean as well as the Spanish territorial seas. An annual plan is drawn up and withdrawal schemes are applied according to EU regulations. The exact difference between the two POs is not clear. Perhaps vessel size was the criterion for dividing them. The vessels in this PO range between 9.6 and 38.7m long (average 28 m), whereas the OPPC3 has vessels ranging from 26.5 and 100.6m length.

Looking at the flow of fish transactions from fisherman to processor, it can be seen that fresh fish and frozen fish follow two different pathways in Vigo. Offshore fishermen, independent owners or skippers working for Pescanova have agents working for them. These agents negotiate the fish price and do the marketing for the fishermen while the latter are still out at sea. The agents have a unique market-price system and publish a select journal for the members. Frozen fish can then be sold directly via a processing company. Often the auction is bypassed. Fresh fish on the other hand, whether caught by inshore or offshore fishermen, is always auctioned before arriving at the final destination. The European consumer can obtain fresh fish in the local market, whether in Vigo or in some other port. However, consumer behaviour is rapidly changing, and often they will buy a fish product that was often processed in another European country and re-imported to the country of origin.

Peterhead, Scotland

The Producers' Organisations in Britain present another interesting example of multi-purpose organisation. As pointed out by Symes and Phillipson in Chapter 4, the authorities allocate sectoral quotas to the POs that their Boards then distribute among members according to principles of their own making. In addition, some POs have more recently assumed a direct role in processing and marketing through bottom up vertical integration. The British case demonstrates the potential of Producers' Organisations to become a more pro-active party in governing catch as well as part of the transactions that involve the whole fish chain. This provides an interesting lesson about the opportunity of extending the co-management scheme beyond the catch side and into the market side of the fish chain.

The public authorities allocate sectoral catch quotas to a PO in January of the following year. This allocation of catch quotas is done according to the aggregate landings of PO membership vessels and vary according to fish species. These catch quotas represent a percentage of total national quota allocated to the United Kingdom by the European Commission. PO membership requirements therefore include a good track record, which is necessary for Government authorities to determine the quantity in terms of catch quotas allocated to POs and the non-sector.

The track record of a vessel comprises the historical individual landings made by a given fishing vessel. In the case of those fishing vessels operating in the demersal fisheries sector, track records are calculated as a total of recorded landings over the previous three years. Those vessels operating in the pelagic fisheries sector have track records based on total recorded landings made over the previous two years (Symes *et al.*, 1996). This system is used to determine the catch quotas allocated by the Scottish Office, Agriculture and Fisheries Department (SOAFD) to the individual fishing vessels comprising the sector and non-sector.

The pelagic fisheries sector is managed on an annual basis, as it is a bulk fisheries and highly seasonal in nature. In doing so, skippers are therefore given maximum freedom to manage their activities how they see fit. Successful quota management strategies also require an element of co-operation between public authorities and POs in order to ensure the long-term viability of the Scottish fishing industry. At the start of each year, the POs are to produce a fishing plan that has to be accepted by the Government authorities and the European Commission. Once this has been completed, the POs are left to their own tools and systems of catch quota management.

In 1994, the British Government introduced legislative measures consenting POs with a right to trade in quotas. This legislative measure follows the realisation of the fishing industry that the added cost of acquiring catch quotas is too great. This makes it difficult for skippers to pay off the debts incurred upon entering the industry. Debts introduce an element of inflexibility within the industry, making it difficult for skippers to diversify their economic activities and/or to form new partnerships. However, it is also difficult for a skipper to leave the industry. Decommissioning funds are only made available with considerable difficulty and with long delays in final payment. Furthermore, the funds offered skippers leaving the industry rarely reflect the true investments made and the potential latent capital captured by the vessel. Hence, in an attempt to help skippers leave the industry, the PO tradings in catch quotas contributes to an industry-funded decommissioning of second- and third-hand fishing vessels. This helps to reduce the total capacity of the British fleet and also provides PO members with additional availability of catch quotas.

Throughout the years, POs have increasingly adopted a commercial role. Initially, POs like the Shetland Fishermen's PO (SFPO) and the Scottish Fishermen's Organisation (SFO), participated in the pelagic fisheries sector by establishing contractual agreements with Russian factory vessels, known as Klondykers. This allowed the Scottish pelagic fisheries sector to develop outlets for the pelagic catch, such as raw herring, in countries like China, Russia and Japan.

Today, several POs also take a commercial role within the white fish market, either in joint projects or in projects they themselves set up. These projects concentrate predominantly on the primary processing of pelagic and shellfish. Some PO projects have also extended into regulating the supply of deep-sea species, such as grenadier. The projects predominantly focus on fish species which are not necessarily in great demand within the UK but do find a market at the international level.

Scottish POs realise that many of these fish species, such as megrims, shellfish, monkfish, witches, turblets and hake, fetch better prices abroad than on national fish markets. Fish landings are more likely to fetch top prices in a semi-processed state as fish quality is maintained and fish products are easier to package and transport.

However, many primary fish processors view the increasing commercial role adopted by POs as having negative repercussions on established processors. They fear POs will take away market opportunities processors would be seeking for themselves. Hence, it is not uncommon for POs to encounter opposition from processors.

POs have only taken a commercial route for a selected number of fish species; that is, those species that POs have found not to fetch a price commensurate with what they consider the market price at the time of the sale. Furthermore, despite this rising commercial approach to fish marketing amongst POs, few actively market their members' fish in ways adopted by the SFPO and SFO. In fact, the SFPO and SFO consider this to be their weakness.

Lerwick, Shetland

The Shetland FPO comprises 60 vessels, nine of which are pelagic boats. The remaining 51 white-fish vessels range from big stern trawlers to small part-time vessels that alternate between shellfish and white fish. The commercial role of the SFPO is crucial to its members. This is because fish markets are highly restricted in places like the Shetlands and Orkney. The restricted nature of markets on the Shetland Islands has forced vessels to land on the mainland in order to fetch profitable prices on the open market. Consequently, the chances of these landings fetching top market prices on the mainland depend on the time it takes for the vessels to reach the shoreline and transport their landings to fish markets.

Fish quality is determined by many factors, such as the storage measures on board of the vessel, gutting procedures, the ice-to-fish ratio and packaging methods. Interior bruising of the raw fish material will bring market prices down, although such bruising is not obvious until the fish is filleted. Furthermore, it is best to land and sell fish as soon as it has been caught; that is, the time the fish remains on the vessel affects the quality of the raw material and consequently determines whether landings fetch good market prices.

With the above in mind, the SFPO has purchased two processing factories responsible for the primary processing of white fish species. This prepares the catch for transport so that less bulk is involved. In addition, the chances of quality deterioration of the fish material are minimised. The SFPO also has a joint venture with a factory on Mainland Britain. This allows the SFPO to actively market the catch of its members throughout the mainland and Continent.

At present, there are two distinct types of fish marketing approaches adopted by the SFPO; notably, passive marketing and pro-active marketing. The passive marketing approach to fisheries management includes the withdrawal price regime. A floor quote on the open market is established below which fish is withdrawn. For example, if a skipper lands on days when there are about 20 thousand boxes of the same catch as his, he will not sell and all of it goes to fishmeal. Consequently, the PO will compensate him. Hence, although not all POs withdraw all fish species which fall below the minimum quote, the PO buys most unsold fish species.

The SFPO, however, holds that it is best to actively manage the raw fish material by preventing the withdrawal from taking place. Hence, rather than allowing the withdrawal to take place and then carrying out the ensuing financial responsibilities, the SFPO buys raw fish material from its members and processes it to customer requirements. In this way, the SFPO creates demand at the secondary processing level, thereby preventing withdrawals. Most of these developments are in the pelagic and shellfish fisheries sector, although the SFPO is looking into the possibility of establishing a similar position within the white-fish sector.

Continuity in supply is achieved by means of constant communication via satellite and radio between the SFPO processing factories and its members. The managers tell skippers when more supplies of pelagics are needed. The skipper is also told when it is best to land; that is, when market prices are most profitable. Fish planning operated by the SFPO is based on finding an alternative for the volumes of fish set to be sold at the best price possible. Furthermore, the SFPO also realises the importance of a constant supply. If no vessel lands any fish, the processing firm will have to send all its employees home but still pay them for a day's work.

Today, the trend of skippers landing directly to processors is common for vessels operating in the pelagic, shellfish and deep-sea fishing sectors. It is not, however, common in the white-fish sector, where auctions dominate the market system. However, direct contract arrangements between skippers and processors are likely to increase as skippers realise that the volumes they can catch and land are more or less fixed. They must therefore plan their landings such that they obtain top market prices. This is a logical course of events as skippers must obtain the largest possible revenue in order to repay bank loans and debts incurred by their business. Skippers are therefore beginning to recognise the fact that they have to play the market. The SFPO hopes to help its members by giving them advice about landing patterns and fish prices. In this way, it hopes to create a market environment that prevents withdrawals from happening

The Netherlands

The Dutch case is an example of a recent innovation in organisational and institutional change within the catch sector. The system described here is known as the 'Biesheuvel' system, after a Dutch ex-Prime Minister who chaired the committee that introduced the system. The Biesheuvel groups are not POs in the formal sense of the concept, but they are closely interrelated to the Dutch POs. In contrast to the British system described above, the catch management functions have been separated, to be handled by a specially designed organisation supervised by the Fish Board, a semi-governmental corporate institution that has existed for many decades in the Netherlands. The Fish Board provides the secretariat of the group and also approves its catch plan. There are no formal ties between the POs and the Biesheuwel groups, but as memberships overlap, there are nevertheless informal ties. The Dutch Biesheuvel system is an example of a well-designed institution that provides incentives for groups of fishermen to assume co-management responsibilities. It also shows that co-management and individual transferable quotas (ITQs) are not mutually exclusive. The Biesheuvel system is presently evolving towards more involvement of trade and processing functions.

The group system was developed in response to continuous problems experienced in the governance of the Dutch fishing industry in earlier years. An individual quota system was introduced in flat fishery in 1976. Throughout the 1980s the Dutch government had problems to enforce this quota system to comply to European catch restrictions. In the early 1990s the regulatory framework was so strict that it was almost impossible, even for well-intentioned fishermen, to comply to the rules and have an economically viable fishing operation at the same time. By requiring fishermen to auction all fish species under quota, the group system improves the transparency of the landing process and therefore provides an opportunity to relax some of the regulations applying to the quota system as well as to the landing procedures. At the same time the group regime intended to involve economic benefits for the individual fisherman. These benefits were supposed to result from improved quota trading possibilities as well as increased market prices. Under the old system of management, the government restricted the quota trade to the early months of the year to simplify enforcement. Under the group regime, trade in quotas between group members was permitted throughout the year, thus allowing an individual fisherman to lease out quota that he could not use himself or to rent extra quota if his own was exhausted. The price improvements were supposed to result from a better match of supply and demand for fish.

For the Dutch flat-fish industry, the major problem is not so much the EU system of intervention prices (for most of the fish landed in the Netherlands, demand exceeds supply), but rather the fact that it is socio-politically unacceptable that the catch limitations agreed in the European context are violated. This is not only a biological problem but also a political issue: since the mid-1980s, the Dutch government has struggled to prevent the quotas allocated to the Netherlands in the context of agreed Total Allowable Catch being exceeded. In former days the predominance of this policy goal could lead to an early closure of the fishing season for all fishermen when illegally landed fish emerging in distributive channels was deduced from the national quota, without recognition from which vessel this fish originated. The consequence of this public action was that fishermen were forced to land more fish than the market could bear in the early months of the year, to prevent them being unable to fish their individual quota later in the year when the season was closed early. This is the 'Prisoners' Dilemma' fishermen faced before Biesheuvel Groups came into play. The threat of early closure of the fishing season for all fishermen not only led to a high supply of flat fish in a period when the quality of the fish was suboptimal (plaice spawns in the first months of the year, the quality of the flesh and the revenue of filleting is low). It also followed that when a closure of the season took place at the end of the year there was no supply anymore, while demand was high (holiday season). The introduction of the Biesheuvel Groups in 1993 explicitly aimed to improve the economic result of fisheries by balancing the pattern of supply with market demand. A more evenly spread supply was expected to lead toward a better price for the fishermen and also to prevent early closure of the season, which meant that traders and processors had to give 'no' for an answer to the supply channel or had to sheer away abroad.

A government-commissioned evaluation study into the group system, conducted in 1996, found that although market prices had not improved as much as expected, there were indeed economic benefits for the fishing industry because the landing of fish was more evenly distributed throughout the year. Another advantage of the group system was that more of the national quota was fished up to the limit granted under European regulations. As a result of improved quota trading possibilities, income was distributed more evenly among fishermen. This was thought to increase the resilience of the industry as a whole. From our own interviews it has become clear that the success of the groups has boosted the confidence of the fishermen that they themselves can indeed solve co-ordination problems in their collective interest through voluntary co-operation. It should be pointed out that the group system, although based on voluntary co-operation, was in fact designed such that few fishermen could afford to refuse entering a group. This shows the important role of government in catalysing organisational innovation. The Dutch government has also construed the group system as an example to other European member states of how government involvement can be institutionalised in the fishing industry along more co-operative lines.

More recently, the government has tried to improve the involvement of the auctions, trading and processing industry in this co-management approach. The subsequent parties in the fish chain were invited to participate by only buying fish that was properly registered and auctioned, rather than sold directly by individual fishermen, outside the group system. This would further improve the transparency of the fish chain, thus making the quota group management system more effective.

The Quality Issue From a Co-Management Perspective

The previous section described the institutionalised efforts to improve the management of the chain from the harvesting side. Zooming in on the role of POs or similar types of Organisations as found throughout European Union, we described the great diversity in the organisational responses to this issue, but pointed out that the system is currently evolving towards greater integration of catch and market. This section will focus on one important governance challenge involving other partners in the chain, where POs currently play a minor role. However, as the quality issue typically involves the entire chain from catch to market, POs may nevertheless makes a more substantial contribution in the future. For our purpose, it suffices to describe how current attempts to impose quality standards are designed and how they are experienced in the (Dutch) fishing industry.

HACCP

With the advent of the EU internal market on the first of January 1993, the conventional quality enforcement system adopted by government authorities throughout the Community did not fit anymore. Product quality was traditionally inspected by means of random sampling of final products. These checks were performed by public authorities. If the product did not satisfy national quality standards, the processing firm in question would be subject to fines. It was obvious to the fishing industry and member states that this system of national inspections was inadequate to meet the requirements of the Common Market. Apart from this concern for harmonisation, it was also thought desirable to shift some responsibility for quality control from the government to the individual companies. Hence, it was decided that the Hazard Analysis Critical Control Point (HACCP) system was to be adopted (Council Directive 91/493/EC).

Under the new rules, fish processors are responsible for continuous product control. The HACCP methodology in fact offers the processor a means to control the quality of fish products in a step-by-step manner, at Critical Control Points (CCPs). In this way, it is not the final product that is ultimately inspected, but the entire production line. The public authorities are no longer responsible for the inspection of product quality; the task is now assigned to employees operating in processing firms. This task used to be to check whether the HACCP system was properly adopted within the firm; the inspector now checks whether 'the quality control system realises set goals'. It is worth noting that the mobilisation of the self-governing capacity of fish processors and traders in quality control provides an example of devolution of management responsibility from government to industry, albeit the micro rather than the meso level. This is still one step short of involving the entire chain. Until that involvement is accomplished, the sector runs the risk of not accomplishing the most optimal performance of quality management. Nevertheless, the new system can be considered an improvement over the earlier situation.

The HACCP system is expected to be more efficient than conventional quality control in that it works on the basis of prevention rather than rejection of defects in output. The HACCP system gives individual processors increased flexibility in that they may interrupt the processing line if they notice deviations from specified quality standards. The HACCP system is cost-efficient to the extent that it provides the processor with a means to analyse, control and possibly reduce the risk of supplying his customers with poor-quality fresh fish and fisheries products. The system comprises the identification of potential hazards and the need to establish criteria which employees must seek to satisfy in order to keep a CCP under control. A monitoring system is enforced accordingly and measures are taken to correct a situation when a CCP is not under control. Procedures of verification as well as documentation and record keeping are established.

The enforcement of the HACCP system throughout the EU has shifted some responsibility for the quality of fresh fish and fisheries products away from public authorities and placed it at the private processing level. Whereas public authorities establish minimum product-quality standards to be applied uniformly throughout the fishing industry as a whole, the private sector will stimulate product diversification by satisfying specific product quality specifications required by individual retailers and supermarket chains. The

HACCP system also provides processing firms with an administrative means towards the prevention of legal claims.

The Dutch Experience

Despite the potential benefits from introducing HACCP, there is a lack of enthusiasm for the introduction of the HACCP system within small fish processing firms. We found that by the end of 1995 there were still large numbers of fish processing firms that did not have a well-functioning HACCP system. This inability or unwillingness of certain firms to adopt HACCP regulation has been attributed to the absence of a quality culture and a lack of professionalism. Firms that already have a well-functioning HACCP system feel that they are in a disadvantaged competitive position vis-à-vis firms that have not (yet) invested in the introduction of a HACCP system. Some advocate, therefore, a more strict public enforcement of HACCP regulation for firms that are 'lagging behind' (Dubbink, van der Schans and van Vliet 1994: 45, 52, 71). From a transaction-cost perspective, however, it is doubtful that the unwillingness of certain firms to conform to HACCP regulation should always be attributable to amateurism and a lack of quality consciousness on behalf of the entrepreneurs concerned. The entrepreneurs interviewed did not reject quality control as such, but doubted that HACCP, more particularly the administrative side of things, would contribute to an efficient and more effective quality control in their particular business context. Furthermore, we found that implementing HACCP is relatively more costly for smaller firms than for larger ones. Small firms often produce artisan-type of products and they supply a local market. It can be questioned whether in a such context there really is a need for *administrative* process control as required by European HACCP regulation. It is quite possible that small fish processing firms do in fact pay attention to quality assurance but that they do not normally organise this through administrative process control but rather through 'mutual adjustment' and 'direct supervision'. Small fish- processing firms often are family-run and, as is commonly accepted in organisation theory (e.g. Mintzberg, 1983) this type of business can be efficiently run through co-ordination mechanisms such as mutual adjustment and direct supervision rather than through administrative process control (a co-ordination mechanism that we may find in the larger bureaucratically organised fish processing firms). The unwillingness of smaller firms to introduce HACCP regulation, therefore, cannot always be attributed to a lack of quality consciousness. Small firms are simply organised differently. For them, the problem is not quality assurance as such, but rather how to organise the administrative process control required in HACCP regulations.

Even if smaller firms indeed do fall short of adequate (informal) quality assurance, it remains questionable that the government should try to enforce HACCP regulation on these firms. Quality control in general requires a change of mentality on the part of the entrepreneur involved and changes of mentality can hardly be brought about through bureaucratic controls. Conventional governmental control directed at minimum product and processing-plant design requirements may be a more effective alternative quality-assurance regime for such firms. This conclusion also places the argument for sector-wide public enforcement of HACCP regulation in a different light. The above analysis suggests that it is not at all clear that the fact that some firms have adopted HACCP while other firms have not, constitutes a case of unfair competition. As long as some sections of the market, in particular the local consumer outlets, accept that some fish processing firms do not have a fully fledged HACCP system, there is no reason why governments should force such system on these firms. To eliminate unfair competition and bring about a level playing field, all a government needs to do is to verify the claims of firms that profess to have an HACCP system. In our view, the government does not, as some have suggested, need to enforce a sector-wide implementation of HACCP.

There is an obvious risk that the HACCP standardisation may contribute to a shake-out of those smaller companies that cannot afford, or do not have much to gain from the administrative-control approach. The HACCP system should be considered a means rather than goal in itself, and there are other ways of improving the quality of fish products than this arrangement. This calls for the pragmatic adaptation of quality-control regulations to specific contexts. It also demands flexibility from regulatory authorities: in one case they may have to expect fully fledged administrative quality control, in another they negotiate adaptations that are sufficiently effective but less explicit than ideally required.

Summary and Conclusion

Institutions do matter, to explain and predict outcomes of economic and social interaction. Institutions contain the capacity not only to restrict and control social interaction, but also to enable and encourage certain social outcomes by structuring the conditions under which people interact. This assumes a perspective of institutions that extends beyond their legal constitution. Institutions consist of rather more than rules and regulations: they also involve the creation of role systems that provide stable expectations and reduce uncertainty in social encounters. Furthermore, institutions sometimes give legitimacy to take on responsibility for collective action such as the provision of common goods. Opportunities are often not realised because institutions to properly deal with them are inadequate or non-existent. For instance, improvements in quality control that would benefit the whole fish chain are not yet accomplished due to this problem. Quality control typically involves the whole chain from catch to market, and yet quality control as an official objective is currently addressed mainly at the individual firm level.

The PO system represents an insufficiently tapped opportunity to strengthen the co-ordination of chain interaction. However, this would require the PO role to be extended from the conventional responsibility of administering the price support scheme. As this Chapter has indicated, POs are well placed to take on additional and more pro-active roles, for instance related to improving the marketing conditions and quality profile of the fish landed by their membership.

There is considerable diversity nowadays in the institutional detail of POs and similar Organisations. This can be explained partly by the historical context within which POs were introduced, and partly by the requirements of the particular fishery. In Spain long-established *cofradias* exist side by side with recently established PO systems to comply to EU regulation. To some extent the two structures have overlapping responsibilities, but the *cofradias* have traditionally performed a much broader scope of functions than the POs. In Britain, POs were given a central role within the sectoral quota management system. More recently, some POs have become directly involved in fish processing and trading in ways that have not been accomplished in other member states. The Dutch

Biesheuvel groups limit their role to improving the co-ordination of catch and market through improved fish planning. Traders and processors have been invited to complement the co-management approach as developed on the catch side into a more comprehensive system of self-regulation extending to the subsequent stages of the fish chain.

All in all, despite the diversity within particular institutional contexts, these organisations are there to fill a gap in the chain between the individual harvester on the one hand and the trade and processing industry on the other. These institutions provide individual fishermen with bargaining power vis-à-vis trading partners that are often much more resourceful. There is no guarantee that quality improvement will automatically benefit the individual fisherman. It is crucial, therefore, that fishermen organise themselves effectively in order to negotiate the terms of trade. This should also benefit the whole chain: the first link has a decisive bearing on the over-all quality of fish products, and therefore must be involved on equitable terms. To the extent that catch control and quality control are systemically interrelated, it is to be expected that these functions will be better co-ordinated within the confines of a single multi-purpose organisation than in multiple single-purpose organisations. This relates to the transaction and organisation costs argument advanced in the theoretical section of this Chapter.

The HACCP system of quality control at the individual firm level is an example of devolved management. In theory, the shift of responsibility from the government to trading partners themselves would improve the extent to which quality control regulation responds to local circumstances. In practice however, there is a risk that if administrative process control as implied by the HACCP methodology is enforced universally, this in fact reduces rather than induces the responsiveness of the system to local circumstances. If quality control were truly based on a co-management approach, it would account for the complexity, diversity and dynamics of organisational solutions in the fish chain and introduce quality control interactively rather than super-imposing an administrative control system on all parties, regardless of their systemic relationships.

Note

1   Based on Contract EU Fair Grant 94/C 185-08; DG XIV, European Union.

7 Experiences and Opportunities

A Governance Analysis of Europe's Fisheries

JAN KOOIMAN

Introduction

Chapter 1 has provided the outlines of a conceptual framework with which to analyse the governance of (European) fisheries (for a more detailed presentation, see Kooiman *et al.*, 1997). The present Chapter will take up this framework in more detail, to show what it takes to create a 'social-political opportunity'. To create an opportunity is to carefully convene actors with a diversity of ideas and interests. They are part of complex constellations of relations and they function within situations with manifold tensions and conflicts. Bringing such actors together calls for forms of interactions in which aspects such as diversity, dynamics and complexity are represented.

Successfully organising such opportunity-creating interactions also depends on the institutional arrangements they are part of. These are not givens, and neither can they be ignored. Part of governing an opportunity-creating process consists of influencing institutional settings such that they are enabling rather than constraining. The diversity, dynamics and complexity of modern fisheries requires both individual and shared responsibilities expressed in social-political mixes of self-, co- and hierarchical modes of governing.

As a result, opportunity-creation in fisheries as a social-political governance issue will be analysed here in terms of four categories of variables:

➤ diversity, dynamics and complexity as general features of fisheries;
➤ opportunities as forms of interactions in which these features are operationalised and made tractable;
➤ first and second order governing variables as organisational and structural conditions within which opportunities are created; and
➤ modes of governing as the mixes of public and private responsibilities for the creation and maintenance of such opportunities.

The present Chapter will illustrate these categories with examples culled from the case studies presented in Part II of this book as well as from Chapter 2. These four categories of variables, it will be shown, are building blocks for an analytical framework with which to gain insight in important aspects of opportunity-creation for fisheries in a European context. On the basis of these insights, the Chapter functions as a 'linking pin' between present experiences with governing fisheries in Europe shown in the foregoing Chapters, and future-oriented considerations of learning and action which will be the subject of the Chapters to follow.

Social-Political Governance of Fisheries

The 'social-political governance' approach to fisheries is a way of looking at patterns of interaction between those involved with fish and fish products and those governing these activities. In many countries recent years have seen a shift away from the public sector towards deregulation and privatisation. The shift from direct quality control by public authorities to indirect control of privately operated quality systems, and the introduction of ITQs show this trend on the European level.

But there are also efforts to shift the balance towards sharing tasks and responsibilities towards doing things together instead of alone, either by the 'state' or by the 'market', aiming to discover other ways of coping with new or existing problems or of creating new opportunities in the conditions and developments in fisheries. As Symes and Phillipson have stated in Chapter 4:

> In the current state of Europe's fisheries, co-management is not an option; it is a necessity. The pooling of resources and expertise in a co-operative system can, in theory, only serve to enhance the quality of management. It may be the most opportune mechanism for coping with the highly complex, dynamic and diverse set of actors and issues within the socio-political context of fisheries governance.

This means that among the three governing modes self-governing (by private actors), co-governing (by private and public actors together) and hierarchical governing (by public parties alone), the 'co' option merits much more systematic attention. From a normative point of view, a particular mix of the three governing modes will often be preferable to placing one's eggs in one of these baskets, with a view to the inherently complex, dynamic and diverse nature of fisheries, which requires governing methods in which all partners have a specific contribution to make.

The 'social-political' approach to governing fisheries calls for ways to analyse its governance in a broad perspective, that is to say, of looking at roles and functions of the different partners in governing, such as fishermen and processors as private-sector actors, but also as citizens; at fishermen's organisations as special interest groups as well as governing networks in which communal and historically developed patterns of interaction converge; at local, regional, national and supranational public bodies, not only as authoritative rule makers but also as partners in private or civic efforts to discover, develop and sustain (new) opportunities for fisheries.

Insights such as these will be argued in the Sections to come. On the basis of the four categories of variables, they will be illustrated with empirical examples from today's European fisheries.

Diversity, Dynamics and Complexity

If fisheries, as a social, economic and cultural phenomenon, is basically complex, dynamic and diverse, this also applies to the way it is administered, controlled, steered and governed. This latter point is often not sufficiently recognised or even ignored by many of the theories, models and analyses used in our field of inquiry. The literature and practice of economically based fisheries management is a good example of this: it basically reduces the complexity of the fishing effort to a one-dimensional *homo economicus*, ignoring the diversity of fisheries and the ways it is governed, paying only lip service to its dynamics.

Diversity

When considering *diversity,* we do not just focus on fishermen as the primary actors in fisheries, but also on all those people involved in the marketing, processing and consuming fish or fish products, and the diversity of all the efforts by public and private actors that take part in the governing of those primary activities in and around fishing. All four cases give examples of diversity as a main characteristic of European fisheries. In Texel and Zeeland, less than 100 miles apart, mussel and oyster growing followed quite different historic trajectories with unique social, economic and cultural consequences. In the acceptance of quality-control measures such as the HACCP systems, large firms are quite keen to introduce them, while small firms are unwilling or unable to put them into operation. Fishermen's Organisations (FOs) in Denmark, the UK and Spain differ hugely in size, organisational form and managerial capabilities. The introduction of salmon farming in its social, economic and environmental consequences showed the different ways this process was 'governed' in Scotland and the Shetlands.

Growing diversity is a characteristic of our time, with its emphasis on individual opportunities and responsibilities at the micro (individuals), meso (organisation and management) and macro (societal) levels. This socio-political given has to be taken seriously. The neglect of diversity as a special and important characteristic of social-political systems is probably at the root of many governance problems. To govern diversity is to influence diverse actors and entities by protecting, maintaining, creating, promoting, and limiting their similarities or dissimilarities. The more diverse the qualities to be governed, the more diverse governing measures are needed, the more decentralisation of governing tasks comes into view.

Countries within Europe differ in style and substance of governing. Denmark has a tradition of centralised and formalised consultation over policy development in fisheries. In the UK such consultations are irregular and informal. In Spain responsibility for consultations on matters in fisheries are decentralised to Autonomous Regions; however, the groups involved exert little influence on the public decision making process on whatever level.

The most important autonomy and diversity-constituting factor in continuously changing situations are the self-images individuals, organisations and groups create. They do so to maintain and protect themselves against the ever-present dangers of dis-integration. Social, economic and political actors will invest a great deal to realise their self-images. If social-political governors ignore those data, this calls for counterforces that are difficult if not impossible to control except by overruling them, with all the negative effects that brings. Insight in diversity can only be gained through actor participation. Acceptance of the diversity of the values, goals and interests as a positive datum is an important element in governing in the modern world. This should be based on openness to a multiplicity of values and trying to understand as well as to analyse them. That this is sometimes forgotten, with detrimental effects, can be shown by the development of the licensing of salmon farming in Scotland in which hardly any consultation took place. This was partly due to the position of the authority responsible for issuing the leases, and partly by its unwillingness to take outside participation seriously. Although there were around 30 interested parties involved, there was no proper framework to get the diversity of viewpoints represented in the lease-issuing procedures. This greatly contributed to the persistence of controversies around the development of the salmon farming industry in Scotland, and to the widespread feeling in public circles as well as private groups that the advent of salmon farming was beyond their control and influence.

Interpreting diversity theoretically as a subject of governing, it is apt to discuss how, on the one hand, the available diversity can be used (created, maintained, supported), where and when it is considered insufficient, while on the other hand it clearly has its boundaries. Social entities cannot function without some measure of uniformity in the inter-human, inter-organisational, inter-regional and inter-state traffic. This is an important issue in fisheries such as the institutionalisation of co-management in which to make FOs part of certain governing regimes. It emerges from the research that FOs reflect a great diversity in interests such as employers *vs.* labour, large vessels *vs.* small boats, pelagic *vs.* demersal fishing and all-embracing or restricted forms of membership. This diversity raises important questions in terms of representation and democratic quality when considered main partners in forms of co-governing. All this implies the need to address questions such as: How can sufficient insight be gained in diversity? How can the degree of preferred or not preferred diversity be determined? Which entities are to be represented? How can the quality of the participating entities/actors be influenced in order for the preferred or necessary similarities or differences to be widened or narrowed down?

Dynamics

Dynamics can be seen as a composite of forces resulting in non-linear or (exceptionally) linear cause-and-effect patterns. Dynamics characterises systems that go from one state or place to another - pushed, drawn or in other ways influenced by natural, technological or social forces. Dynamics therefore can be considered as the change potential of systems. This potential always consists of two opposite forces: those pursuing change in existing situations and those trying to keep things as they are. Dynamics is expressed in the tensions between these two forces. I think the question of the 'why' of dynamics can best be answered by the theory of irreversible processes (Prigogine and Stengers, 1988). This theory entails that open systems such as social systems, because of the exchange of energy / information with their environment, usually are not stable or in equilibrium; they fluctuate constantly. These fluctuations by means of positive feedback processes can take on a 'dramatic' character: they become 'irreversible'. This can especially occur when systems are highly disequilibrated.

Why is this principle of non-reversibility so important? In this perspective, non-linearity is the rule, linearity exception in open systems while dynamics will be expressed in 'jumps' rather than stepwise and irreversible processes can be seen as sources of dynamic order.

The historical development of Texelian fisheries is an interesting source here. Fisheries there became important from the 11th Century on, but due to silting and other environmental factors it declined in the 16th Century. Because of the rise of Dutch maritime commerce, new opportunities arose and many Texelians became pilots or seamen. They also became successful oysterers. The French occupation put and end to this but modernisation brought fisheries back to Texel. The recession of the mid-1890s dealt fisheries a severe blow. Oyster yields also grew, but a rapid expansion of climatic factors exhausted the parks. Many oyster fishermen switched to exploiting eelgrass and discover niches in the market. However market expansion also meant lengthening the market chain, making fishermen more dependent on international competition and business cycles beyond their control. Many turned to shrimping or started to catch flat fish in the North Sea. This changing behaviour over time can be seen as adaptation to dynamic environmental factors, market situations and as a consequence of their own behaviour. Not only the irreversibility, but also the non-linear dynamics of fisheries over time comes out quite clearly in this example.

Governing the dynamics of social-political systems can draw on well-known cybernetic ideas and concepts such as positive and negative feedback. What kinds of social-political tensions can we expect to create dynamic cybernetic qualities such as positive or negative feedback loops? How are these loops organised and can they be influenced?

The farmed salmon market is a good example for discussing (the problematic aspects of) positive feedback. As long as the price for salmon remains high, production is increased. Price levels will drop but supply cannot immediately adapt. It takes several years to grow marketable salmon. After an unprofitable period, profitability increases, and a similar cycle starts all over again.

An example of negative feedback is when mediation has a dampening effect in a conflict that may easily run out of control. This can be seen in the way health agreements in the Shetland salmon industry came about. Thanks to the intermediary role of the local authorities, systematic and monitored patterns of communication on a voluntary basis were established between salmon farmers using the same loch.

If social-political governing is to take the dynamics of modern fisheries seriously, it also needs to take the dynamics of its problems and opportunities and the dynamics of acting on them seriously. This means looking for tensions in social-political interaction patterns. Dynamic governing modes should be applied to these areas of tension. The co-management debate is important in this respect. If, according to Gislason (1994), 'the objectives of fisheries management are now gradually changing from the sustainable use of the commercially valuable resources to the conservation of marine environmental quality', this is likely to have significant implications for fisheries management throughout Europe. It addresses the issues of the tensions and integration of different sciences, resource user groups, environmental organisations and state administration. The dynamic forces inherent in this new definition of issues, participation and representation might mean a completely different way of governing fisheries issues, and the dynamics of this new situation might be a creative force, not only phrasing in new questions, but also in finding new solutions. In other words: this emerging new force might be the creative force for new and unexpected opportunities in fisheries governance.

To make issues such as these explicit is to recognise dynamic potentials. Tensions, present or latent, can be defined as loops having open and closed elements, short or long cycles, diminishing or reinforcing effects, renewable and non-renewable qualities. Recognising this is the beginning of the development of conceptual and practical notions on the use of dynamics in ways of governing.

Complexity

The development of the Western world is predominantly characterised by differentiation and specialisation. The sciences and bureaucracies are great examples of this. The consequence has been great progress in the knowledge of parts, but much less so in the understanding of their interrelationships and the relations between parts and wholes. Complexity seen as the number, quality and intensity of relations between parts and wholes is an important element in a large number of governability questions. Handling complexity in terms of governing the many interdependencies between and within systems (steering, guiding, managing, directing, and controlling) confronts us with a number of rather difficult conceptual and practical questions.

However frequently complexity is identified as an aspect of modern societies, it does not make it easier to define. A good start is to look at the number, intensity and overlap of relations between elements of a system. In fisheries, then, the fisherman and his work are more than a purely economic given: fishing and all the activities following and related to it can be seen as a network of interdependent activities. This is made particularly clear and described in detail in the way fisheries-as-a-chain is run and managed in Scotland. Pelagic and demersal fishing, processing and marketing are closely interrelated as multi-faceted chains and patterns of technical, economic and social interactions, and as a result constitute systems of relatively great complexity.

We assume that there are boundaries to the human capacities to know and to act. In coping with complexity we have to follow the path of combined strategies, e.g. reducing it without losing sight of the essence and of the boundaries of the subjects we deal with intellectually, emotionally and practically.

Not only the practitioners in the field of governing fisheries have to find some way of dealing with the complexities of their situations; this also applies to scholars who try to come to grips with a particular subject matter. This study in itself can be seen as an exercise in coping with complexity. The analysis and presentation of governing fisheries itself is intended as a process of decomposing – (presenting case studies) and composing (integrating these insights in this and following Chapters).

An interesting case of ignoring complexity at least partially is the way Gareth Hardin phrased his well-know parable of the Tragedy of the Commons. As shown in the Salmon study, the way the Crown Estate and River Purification Boards in Scotland interpreted their role in preventing the spreading of epidemics was at odds with the solution advocated by Hardin: formal property rights, bureaucratic controls and/or internalising monetary cost. Hardin overlooked the fact that owners of a resource may have other goals than sustainable exploitation of the resource (e.g. short-term profit, maximum employment). He also did not provide a solution for the fact that even if owners are prepared to manage the resource in a responsible way, they simply may not have the appropriate knowledge: one just had to learn by doing. Learning by doing presupposes processes of communication and information exchange between parties involved, beyond the level conceived of in Hardin's perspective.

Dealing with the complexity of social-political phenomena can be seen as a reasoned and controllable way of analysing and synthesising. If we see complexity as many parts with many relations between them, we have to deal with the parts, the wholes and the relations between parts and wholes. Reduction has to take place, but where and how? At the very least there should be an overview of potential interrelations involved. These interrelations can be substantive, such as economic, social and cultural ones, but also more formal such as in constitutional arrangements or in administrative co-ordinating procedures. Whatever these may be, it seems to be essential in coping with the complexity of social-political issues that this composing and decomposing process of considering effects for parts and wholes be open and accountable, and that criteria with which interrelations are taken into consideration and which can be left out are debatable and clarifiable. In this context, it seems, public organisations have a special role to play in governing complexity. Public authorities are best placed to organise such composing-decomposing 'games' (De Leeuw, 1986) and basically they should be the ones to design and protect such criteria-setting and application procedures.

Conclusions

These are matters of an institutional nature, as will be clear from looking into the issue of co-management and devolution of responsibilities for aspects of fisheries according to principles of subsidiarity. The composition and decomposition of what and who is involved varies among the different levels involved. At the EU level, the definition of the fisheries system will include the political relation with East-European countries, and therefore the consequences of including or excluding 'Klondykers'; for the fisheries chains this is mainly a matter of commercial interest, while for the national governments involved in governing the resource there are environmental, social and economic interdependencies at stake. This requires combined structural and procedural arrangements as a necessary condition to enable those involved in fisheries in the public and in the private sector to handle the complexities at the different levels of responsibility and to cope with those complexities in a responsible manner.

Considering opportunity-creation for fisheries as a 'situation to be governed', it an be concluded from the analysis that the following aspects merit particular recognition:

> *The great diversity of (European) fisheries is one of its most basic assets, and has been insufficiently recognised in the governance of opportunity-creating processes.*

> *There is sufficient externally induced dynamics within (European) fisheries to be used in the governance of opportunity-creation as a creative force rather than as an untamed one.*

> *Governing fisheries (in Europe) is particularly hampered by the lack of theoretically based notions on its complexity, which makes it very difficult for those governing opportunity-creating situations to design and apply proper (de)composition procedures.*

Interactions

As indicated in Chapter 1, interactions play an important role. But in our analytical framework they serve as an important medium to try to come to grips with the diversity, dynamics and complexity of socio-political phenomena such as fisheries and its governing. We regard interactions as interrelations between actors in which something new is created. An interaction is more than an exchange or a bargaining process. It is also more than an incidental meeting or an accidental occasion. There is an element of mutuality: an interaction affects and influences the partners involved. There is an element of interdependency in an interaction and, to some extent, the actors involved interpenetrate each other's domains. These interdependencies, the new or creative element in them, and the different forms these interrelations take, will be discussed in the following Sections. Here I shall deal with three elements to be distinguished in interactions: the *images, instruments* and *action* (*potential*) which the actors bring into their interactions. These three elements take place at a direct or intentional actor level. However they are also embedded within more structural contexts in which the interactions (always) take place: images as parts of *culture*, instruments as parts of *resources*, action (potential) as part of *power* distributions.

Images

Images play an all too often underestimated role in all social-political governance – including fisheries. Governing cannot do without the formation of images. Everyone governing in whatever capacity or authority does form ideas as to the how and why of that governing. The question is not so much if there are such ideas or images, but how explicit and systematic they are. Images can be extensive and based upon thorough analysis, but also limited. They are informed by the private experiences of the governor; they can be clear or conscious, but also hover in the background. All sorts of images can be distinguished: visions, knowledge, facts, judgements, presuppositions, hypotheses, theories, convictions, ends and goals. Image formation about social-political issues such as in fisheries often contain complete or partially implicit images of man and society, in which existing and preferred situations are mixed. Governors in the political arena, but also those who govern in social settings are just as prejudice-ridden as the next man about what is good or bad in man and society, about social and political problems and opportunities. In the reality of modern governance images will differ on almost any issue: some principled, others more pragmatic.

A set of such images plays a crucial role in the annual decision-making on quotas in European fisheries. The mix of scientific-biological estimates, political-administrative considerations and vague and often unsubstantiated ideas on fishermen's behaviour are the decisive factors on which key instruments in governing European fisheries are based. The great difficulties and lack of trust in the interactions between fishermen, their organisations, environmental organisations and the public or semi-public authorities involved in regulating European fisheries are to a large extent due to this mix of partly founded, partly constructed and partly ideological images.

From what went before it will be clear that these images are embedded in broader patterns. These are akin to what the literature considers 'culture': general systems of beliefs, values and norms that are commonly shared among a group. Predominant images in a certain pattern of interaction can be in line with - or at odds with - more broadly shared ideas, norms or values. This might be an important indication of a tension that might lead to a non-linear dynamic action or reaction.

How this applies to fisheries in a historical context can be see in the study on the mussel and oyster industry in Texel, the Netherlands. As in so many fishing communities, local social relations are characterised by an egalitarian façade, though actual relations are certainly not egalitarian. And the egalitarianism of the Yerseke fishermen is ideological. This became quite clear in a conflict in the 1950s around the drawing of lots for mussel plots in the Wadden Sea. The conflict took on an ideological character and divided the town. Old friends fell out with each other, brothers and cousins became rivals or even enemies. Crewmembers of discordant planters were not allowed to sit at the same table in the taverns they frequented. Two associations were formed, which usually met on the same day, often in the same building - if in separate rooms - and discussed the same topics.

What must the concept of 'image' be able to do? For our purpose, the most important contribution of conceptualising images and image-formation processes is to construct a kind of 'theory-in-practice'. This structures observations and accompanies image formation processes as a kind of manual, helping to create effective and legitimate images for governing. The attention we pay to *opportunities* as coequal to *problems* is a good example here. Images are always there, but they can be created and modified. The formation of Biesheuvel groups in the Netherlands, which are partly organised on a regional basis is another interesting example of the use of 'theory-in-practice'. The groups are responsible for the division and control of allotted quotas of flat fish. The underlying 'theory-in-practice' was the Prisoner's Dilemma: as each fisherman fears that his colleagues will overfish their allotted quotas, he will be quick to exhaust his own quota. By means of a system of carrots and sticks, the fishermen were coaxed into forming such groups to overcome individual Prisoner's Dilemmas (Dubbink and Van Vliet, 1996).

The importance of explicit and systematic image formation in governing fisheries is manifold. Uncovering images can show the limitations of dealing with diversity, complexity and dynamics of governing situations in fisheries. Secondly, explication of images underpins governing legitimatacy. By continuously checking images, the processes in which they are formed can be brought under control and be open to criticism. This is the basis for their potential acceptability in the democratic governance of fisheries.

Instruments

The importance of conceptualising the instrumental condition of governing lies in the almost unlimited range of alternative tools societies have at their disposal for governing themselves. Governments as well as private or semi-private actors have a choice of tools at their disposal. Some public instruments find a wide application (information, taxation) and involve almost everybody, others are quite specific, and directed at a single individual or one enterprise (a grant or a license). Some have a very formal quality (a law or a treaty), others are quite informal (convincing, paying a visit or making a speech). But social actors also have an enormous range of instruments at their disposal: varying from organising themselves, demonstrations (protest) to solidarity actions and voluntary expertise (support).

Licenses for salmon farming in Mainland Scotland and Shetland show many of the aspects of the application of a public instrument and their institutional framework. In Scotland the Crown Estate, formally an independent body, shows almost all the characteristics of a single objective body: little consultation, no feedback, little consideration for such broader issues as industrial development and other interests. In the Shetlands, the Shetland Island Council has unique powers to regulate all developments in its coastal waters. The SIC issues all licenses in the Shetlands and controls all the conditions attached. This has worked out in practice in broad consultation, public debate in a democratically controlled body, regular planning procedures with all guarantees attached, consideration of other interests, licensing part of industrial and social development, conflicting objectives politically weighed and legitimised. The license as a tool of governing in those two situations shows not only what it can do, but also the resource base it is part of.

Public and private 'governors' need to make choices between instruments or combinations of instruments to be used: rules, law, planning, finances, organisation, information, threats or promises. More theoretically phrased: instruments with positive or negative incentives attached; with positive or negative feedback qualities; with a feed-forward, feed-while or feedback character; with a generalising effect or only a specific 'audience'.

As the study on mussels and oysters has shown, a combination of instruments can be used to control and steer different aspects of this industry; it also shows their changes over time and flexibility in application. Instruments such as licenses, financial support, bans on the use of certain shells and yearling oysters, supply of roofing tiles and credit, access permits, price regulations, setting quality standards, allocating production quotas (standard capacity numbers), allotment of plots were used in changing combinations and degrees of severity.

Instruments, separately or combined, influence social-political governing or conversely, governing patterns influence the use and effects of certain (combinations of) instruments. A number of interesting and important questions can be raised here. How are instruments developed, what kinds of tensions, conflicts characterise their choice processes? How effective is a certain instrument? Who profits and who doesn't, who will be punished and who won't? What kind of side effects can a certain instrument have and how serious are these?

Often instruments need to be developed to satisfy divergent demands. For that reason one has to accept a suboptimal outcome, that is to say, be content with a compromise or drop one or more demands on a certain instrument. An alternative would be to combine different instruments, but these have to be adjusted to each other and their sometimes negative aspects reinforce each other. There is also the difficulty of getting rid of an instrument when necessary, for example when a problem situation has changed, and opportunity materialised, better instruments have been developed, or other perspectives on problems have come to life. The poor functioning of instruments will usually be absorbed by the application of other, supplementary instruments.

Action

Interactions do not take place in an a-political world. Action (potential) is also expressed in the structural context, in which the close relation between action (potential) and power bases are introduced in the analysis. Again, all the research projects show the importance of this element. Not only we find the more obvious action potential and power base such as those of the Dutch government in setting up Biesheuvel groups after a political crisis about the lack of control of quotas (Dubbink and Van Vliet, 1996). We also call attention to the growing action potential of supermarkets and their chains in setting the standards for the price, quality and distribution of fish and fish products. The management potential (or the lack of it) of FOs to become major players in fisheries chains is an example of what we mean by the action or power component of social-political governance. Another action role is the ability of environmental groups to influence fisheries towards a more sustainable use of the resource. There is question here of the action potential of 'civil society' as a major socio-political institution in relation to the state and the market.

In public as well as in private sectors of society, enormous amounts of action and reaction potential are piled up. This is part of a pluralistic form of society where the sources of power and influence are divided among many actors. Action and power are embedded in social conditions and simultaneously shape them. Not all sources of power lead towards governing action or reaction, and certainly not all potential actions are carried through. Sometimes the dynamics in governing situations will take care of the action potential being realised; sometimes their complexity or diversity will prevent actions being taken. That is not done at random, though: there are observable patterns in these processes. There are situational and contingent factors, but also more or less permanent conditions that promote or restrain governing action. In economically and socially deprived regions there may be no other choice but to exploit the natural resources that are locally available. So it can be argued that local councillors in Scotland, whether or not they had the formal power to prevent the development of fish farming, actually had very little leeway for action for lack of other means of employment in the region. Thus, prospective fish farmers, especially in the early days of development, were in a strong bargaining position vis-à-vis local community representatives.

Formal decisions can be seen as a *caesura* in action-reaction processes. Chains of governing interactions are interrupted for a moment; a state of affairs is formalised by taking a decision. Thereupon governing interactions often assume a different character; other actors may enter the arena, the dynamics of action and reaction often changes character.

This can be seen quite clearly in the way the EU's Council of Ministers takes formal decisions separate from fisheries policy making, execution and implementation. These latter actions are the responsibilities of the Member States, which basically enjoy the freedom to implement European rules in their own way within the framework of common decisions. A completely different set of actors, with different interests, motives and power bases comes into play here. This means that people at the end of the line, such as fishermen or other market actors in different European countries within the context of one policy are confronted by completely different sets of implementation regimes. They not only consider this confusing, but also inequitable and even detrimental to a proper operational management of their business.

In democratic polities, public governing often is reactive rather than pro-active. Social and political governing bodies often lag behind developments not only because of bureaucratic inertia, but also because of the majority principle in collective decision-making. More innovative action can be expected at local or regional governing levels. That calls forth the question under what types of conditions action will or should be activated. This is an important element especially when we are talking about creating (new) opportunities for fisheries. Making an opportunity succeed is very much bound up with the organisational capacity to set an action-reaction chain in motion, aimed at reaching a certain amount of positive feedback needed to create a sufficient amount of social, political and administrative willpower. This is very much a 'systemic' quality that does not seem to have developed sufficiently in fisheries.

Conclusion

Interactions are the basic stuff of governing. Without democratically organised and substantively legitimated interaction patterns between those governing fisheries and those governed, its social-political governance can hardly be effective. From the analysis it may be concluded that the main patterns of interaction in European fisheries have insufficient mutually 'interpenetrating' effect. A main factor seems to be the mutual lack of trust, based upon a lack of endurance in those interactions. Broken down into the three elements, images, instruments and action, the following statements may be made about those interactions:

➢ *In fisheries the available governing images are too one-sidedly aimed at preserving the resource by limiting the catch, to be of much value for opportunity-creation for fisheries as a whole.*

➢ *Developing proper social-political instruments for opportunity-creation in (European) fisheries needs a fresh start, instead of expecting much result from adding more instruments on top of the existing sets.*

➢ *The action-related efforts that might have the greatest impact in creating new opportunities in (European) fisheries may be strengthening the links between them.*

Orders of Governing

In the analysis I draw a distinction between *first* and *second-order* governing. First-order governing consists of governing activities aimed at solving problems or creating opportunities directly at a particular level of aggregation and a social-political situation in particular. Second-order governing consists of governing activities aimed at creating or changing the conditions under which first-order problem-solving or opportunity-creation takes place. In fisheries, setting quota in order to limit over-harvesting of fish is first-order governing; changing the institutional EU framework within which measures such as quota setting takes place is second-order governing.

Continuing the earlier distinction between the intentional level of interactions and the structural one, first-order governing interactions can be said to have the direct intention of solving a governing problem or creating a governing opportunity. Second-order governing is about influencing structural conditions under which these intentional interactions take place. Or to say it still a little bit different: images, instruments and action (potentials) are the key elements of first-order governing; second-order governing focusses its attention at cultural, resource and power patterns.

While first-order governing can be seen as balancing processes between problems, solutions and opportunity-creation, second-order governing may be conceptualised as a balancing process between needs and capacities. These are more than terminological niceties; they express that first and second-order governing have a different character. In systems terms first-order governing can be said to focus on parts of a system, whereas the system as such is a constant; in second-order governing the focus shifts to the requirements of the system as a whole, while its parts are held constant.

As a third, superimposed order, meta-governing is identified where governing as such is the focus. Relevant questions include: who or what governs the governors, what are the basic qualities that governing-as-such has to fulfil? These are questions of 'governability' that we can try to answer in rather specific terms for a sector like fisheries, but they also apply to more general aspects of modern liberal-democratic societies. We shall return to these metagoverning aspects in the final Section of this Chapter.

First-Order Governing

This Section will develop some ideas on social-political problem-solving and opportunity-creation as the main first-order governing activities. This is a useful distinction, not only because these two activities represent the main challenges in modern societies but also because the two types of activities are not the same from a conceptual point of view. They require different sets of social-political interactions, with different patterns and different outcomes.

Problem-solving in diverse, dynamic and complex situations goes through a cyclical process including a number of stages in which images are related to instruments, and instruments to actions. In the process of image formation the process, or rather the cycle, runs from identification of those who experience problems (dealing with diversity), through stocktaking of the interactions taking part in terms of partial problems or problem aspects (dealing with complexity), to the localisation of pockets of tensions in interactions which can be identified as sources of problems (dealing with dynamics). This process ends, in principle, only when no new discoveries are made and the 'problem space' has more or less been defined. Then the 'solution space' can be defined in which the same sorts of interactions may take place but now focussed on (potential) solutions. This part of the total process can also start earlier, to be part of the problem definition stage. The idea is that the diversity, complexity and dynamics of a 'problem-solution space' are handled in a number steps or phases which are open to scrutiny, discussion and accountability of and for those involved in it - on the governing as well as the governed side.

Handling the diversity, complexity and dynamics of opportunity-creation is somewhat different. There usually are no experiences to be taken stock of and identified yet. It is often an individual or group, either on the governing or the governed side, that spots an opportunity. Opportunities do not start from the diversity of those participating in a situation, but rather from the experience of dynamics within a certain situation. An opportunity can be said to be a positively experienced tension to be evaluated from a future-orientated perspective, while a problem is more like a negatively experienced tension within a past-oriented perspective. What are the relevant tensions that bring on the opportunity experience? Which are the parts of the system that might be effected by the opportunity: which types of interactions are potentially involved in the opportunity? Then the actors who might participate in the necessary interactions and instruments and action potential are identified. If the image, instrument and action potential is explored, the opportunity space can be defined and its boundaries drawn.

In a stylised way some of the dynamics, diversity and complexity aspects of the introduction of salmon farming in mainland Scotland and the Shetlands can be analysed in terms of the first-order governing of an opportunity for those areas.

Dynamics play an important role in the creation and use of an opportunity. In Scotland the actors or actor groups involved seem to have been unable to create an 'opportunity space' out of the advent of salmon farming in their region. The advent of this important new economic activity was mainly a process of tensions without proper outlets, which meant that not even a proper opportunity space was defined. Particularly at the start, it seemed a missed opportunity, which was only mended at a later stage. The introduction of salmon farming in the Shetlands can be explained much better in terms of a properly defined opportunity, where the dynamics and the complexity (and with some hesitation also the diversity of interests) of the situation were understood, the proper instruments applied and sufficient action potential was available to reap many of its benefits.

The other case studies also provide examples of ways in which problems in European fisheries are defined, solutions tried or found, opportunities defined and used. Some of such problems might also be defined as an opportunity: a problem for one actor might be an opportunity for another. Likewise, HACCP is seen as a problem for smaller processors, but an opportunity for larger ones. This is attributable to the multi-faceted character of first-order governing in fisheries, but also to the fact that neither problems nor opportunities are social-political givens, but agreed or disagreed-on inter-subjective notions between the main actor groups involved. A large part of first-order governing consists of finding effective and legitimate inter-subjective agreements of what a problem or opportunity might be. In fisheries there seems to be at least partial agreement among the major parties involved about the nature of the key problems; this seems less true of its opportunities. This could be attributable to a dominant focus on problems and the absence of a search for opportunities. More systematic attention to the different aspects of first-order governing could help to shift this balance.

Second-Order Governing

Besides problem-solving and opportunity-creation, social-political governing includes the responsibility for thinking about creating, sustaining, changing or ending conditions under which first-order governing takes place. It would be far too easy to assume that problems are solved or opportunities created according to criteria and norms as formulated in the former Section. Governing fisheries is hardly an exception in this respect. The plea for new forms of management in fisheries is a case in point.

The systematic look at the conditions under which first-order governing takes place can also be phrased in terms of problem-solving at a higher level. To analyse this higher level, i.e. second-order governing, we shall introduce two other concepts here: governing needs and governing capacities. Governing needs concern problems with first-order governing and capacities with which to address these problems with first-order governing. In analogy to the way we concerned ourselves with first-order governing, second order governing can be considered as a mode of handling 'need-capacity systems' (balancing). A working definition of second-order governing could be: the balancing process between governing needs and governing capacities of a particular social-political system. An extremely good occasion for this balancing process in European fisheries is the preparation of a new Common Fisheries Policy. It depends very much on the analytical and political capacity of those involved if the right level of specificity (1st-order) and generality (2nd-order governing) of questions and answers will be formulated, mixed and integrated in this new policy or, as we prefer, governing perspective.

The patterns of second-order governing of fisheries in Europe have weak and strong links. There is not much of a 'countervailing' power to the overwhelming structural dominance of political and (public) bureaucratic organisations in the governing of European fisheries. In the annual setting of TACs, nationally organised fishing interests seem to be able to resist what are in their view punitive catch limitations. At the European level, however, there are few indications that they play a great role either in setting new goals for fisheries or in creating new opportunities. Neither FOs at the national or international level nor organisations with a civil-society character representing non-fishing interests seem to be able to muster much institutional capacity in response to the governing needs these fisheries display in many respects.

The fishing industries represent a very loosely structured, unstable and often discordant coalition of interests which not only reflect differences between the harvesting and post-harvesting sectors but also their fragmentation according to particular structural and regional preferences. Yet there appears to be a potentially significant relationship between the general political culture and the organisational structure of the fishing industry. Where the industry sits at the negotiating table, in Denmark and in the Netherlands, there is a tendency for stronger and more coherent representative organisations to emerge. Where, on the other hand, representative organisations in Spain and the UK compete for the attention of government departments through informal consultation, fragmentation and conflict of opinion may be more frequent.

In order to systematise our considerations on second-order governing we need an analytical framework. Isn't the framework we applied to first-order governing also useful for second-order governing? Can the different stages we developed for coping with (or analysing) the diversity, complexity and dynamics of social-political problem and opportunity situations also be used for second-order problem and opportunity situations? Comparable principles seem to be applicable; but differences will arise, too. In first-order problem-solving or opportunity-creation, the main 'system' governing perspective emanates from a part, runs through considerations for the whole, and comes back to the part in question. In second-order governing, when a need is formulated, the main perspective is the whole. During the process considerations are focussed on a part or a number of parts in terms of needs-capacities of those parts. But eventually, the perspective will again have to be holistic.

Conclusion

Calling attention to orders of governing in fisheries is informed by the realisation that its governing is a multi-layered activity: not so much in the sense of organisational or constitutional levels, but in the sense of different types or orders of activities – sometimes by different actors, sometimes by the same actors, which together form the governance building. On the one hand, there is concrete problem-solving such as over-fishing and opportunity-creation such as developing new markets for fish products; on the other hand there is the care for the institutional context in which these primary governing activities take place. Focussing on opportunity-creation, this may be phrased as follows:

➢ *A shift in attention away from problem-solving directed to opportunity-creation seems to be a prime condition in creating such opportunities;*
➢ *An important element in creating new opportunities in (European) fisheries is strengthening the inducement to co-operate at all levels, but in particular at the lower one;*
➢ *Governing by stimulation and regulation (carrot and stick) seems to be a promising way to stimulate opportunity-creation in (European) fisheries.*

Modes of Governing

Traditionally public governing has predominantly been associated with state intervention. But from the perspective of social-political governance, it seems profitable to broaden this view. Not only because other forms of governing can be distinguished, such as co-arrangements in which public and private partners jointly take responsibility for a governing task, such as co-management schemes in fisheries. The concept of self-regulation or self-governing is on the rise as well. Another reason for broadening the concept is that it enables us to call attention to all kinds of social-political governing such as by churches, voluntary and professional organisations, and action groups currently typified as civil-society or third-sector actors.

My conceptualisation identifies three modes of governance: self-governing, co-governing (or interactive governing) and hierarchical governing. In the late 1980s and in early 1990s, self-governing became ideologically popular, while interactive or network governing has been receiving more attention in the literature. In the mid-Nineties there seems to be a comeback of hierarchical forms such as the perceived need for a limited but strong state. All three modes of governance merit some discussion.

Self-Governing

Primary social entities such as families, small groups and organisations govern themselves, but larger social and international entities such as churches, the Red Cross and multi-national business firms are basically self-governing bodies as well. The idea of self-governing has the strange connotation of powers being 'handed down' to them by the State. This makes discussions about so called self-governing tendencies of sectors of the society a little odd. Of course they are all subject to official rules and regulations, but in their core activities they are self-governing units and actors. From a social-political governance point of view this is an enormous self-governing capacity. The case study on the FOs shows a rather diverse self-governing capacity of these organisations. It is also clear, however, that partly due to local, and regional circumstances, POs and FOs such as in Vigo, Spain and on the Shetlands have become active self-governing actors within the chains they are part of.

The strange use of the concept of self-governance may also be contingent on the lack of scholarly attention to what self-governing might be. Since a few years, the concept of *autopoiesis* has been on the rise, a concept very close to self-organisation. This concept, originating in biology, can be considered controversial, nevertheless it offers some entries into the way social and social-political systems govern themselves. What is important for our purpose is that this theory tries to understand why societal sectors such as law, economics and science but also public and private bureaucracies seem to develop increasingly independently or autonomously. They ignore, as it were, the broader societal contexts they are part of: they seem to obey developmental 'laws' of their own, they seem to be 'autopoietic', self-organising. They select from their broader environments only what they need and increasingly interpret outside signals (information and interventions) in the application of their own self-images. According to these theoretical notions it is not the 'objective' qualities of governing acts that are of primary importance. Only if and when social-political systems have internalised external governing messages can they reap some effect. Otherwise they will either be ignored, limited or exaggerated. In this sense many a fisherman can be looked upon as an 'autopoietic' actor.

What makes self-governing an interesting phenomenon is, in the first place, that it is already there, and to a much larger extent than is often recognised. So exploring its characteristics seems to be a profitable exercise. Utilising self-governing capacities as part of mixes with other governing modes, it seems to be a good strategy to know at least its strengths and weaknesses. Defining governing not so much in terms of structures but as on-going interaction processes, self-governing capacities can be much easier be discerned. So, there is no structural arrangement governing the market for, say, salmon, but there certainly are social, economic and cultural 'laws' that govern these markets pretty clearly. The diversity, dynamics and complexity of these laws may not be known in every detail. However we know that certain aspects have self-governing capacities, while others have not. Using this knowledge more explicitly may become important as data for self-governing systems themselves, but also for other forms of governing, such as in co- and hierarchical governing mixes. In the governing of fisheries this may become one of the key issues in the years to come.

Interactive or Co –Governing

There can be hardly any doubt that the emergence of mixed forms of co-operation between public and private actors is a function of broad social developments. Growing social differentiation also engenders increasing dependencies. In this context, the emergence of policy networks is considered an important change in the political decision-making structure. Others consider the development of mixed public-private forms of interaction in terms of the need to solve social-political problems, in line with our own pre-occupation. Public-private partnership, a specific form of social-political governance, has been at the centre of interest for some years. The growing interest in co-operation between public and private parties has been informed by economic, social, political and cultural changes. As a consequence of this, it is increasingly asked whether certain issues could not be dealt with more effectively and efficiently by joint action by public and private parties, rather than their acting separately.

To co-govern is to make use of organised forms of interactions for governing purposes. In social-political governing, these are key forms of 'horizontal' governing: parties co-operate, co-ordinate and communicate without a central or dominating governing actor. It is in particular these forms of governing that are better equipped than other modes in diverse, dynamic and complex situations. An interesting form of co-arrangement found in fisheries is known as co-management. Two applications of interactive governing stand out: communicative governing and co-management.

New patterns of governance are addressed to stimulate learning processes that will lead to co-operative behaviour and mutual adjustment, so that all or most actors involved share responsibility for managing structural changes. One such alternative form of governance based on the image of complex problems in which problem-resolving capacities are distributed across autonomous but interdependent actors, is described as communicative governing. In this type of governing a form of rationality is presented in which social actors are considered 'reasonable citizens'. This is a different kind of rationality from the selfish, opportunistic profit or benefit-maximising type underlying economic or public-choice theory. This call on 'reasonable citizens' corresponds with the concept of communicative rationality that is considered appropriate in complex problem-solving as a substitution for instrumental, functional or strategic forms of rationality. The concept of communicative rationality is based on the interaction between human beings; truth and knowledge are based on a (temporary) inter-subjective understanding. In an 'ideal dialogue situation', the only remaining authority is the 'sound argument', which can be related to empirical as well as normative judgements (van Vliet, 1993).

Co-management, by definition, means that government agencies and fishermen, share responsibility for resource management functions (Jentoft, 1989). Co-management, according to Jentoft, tries to steer a middle course between government regulation and community-initiated regulation. Co-management is not as informal as community-based management, it requires fishermen to establish organisations with formal leadership and an executive staff. But this leadership is participatory rather than hierarchical, and (where feasible) decentralised rather than centralised. In these aspects co-management differs from management by government agencies. Co-management is more than consultation: fishermen organisations not only have a say in the decisionmaking process but also have the authority to make and implement regulatory decisions on their own.

The case study on FOs (the most complete in this field to date) puts the picture into perspective. On the basis of a study in Spain, Denmark and the

United Kingdom, it is clear that the structures that are currently in place are more or less suited to the particular tasks of representation and a limited range of devolved management responsibilities. Not all the constituent organisations share the same level of capability in terms of finance, leadership, management skills and aspirations. Indeed, the analysis has indicated that many small, local organisations are unable to acquit themselves adequately of their responsibilities for want of sufficient financial and administrative resources. While the industry is indeed overpopulated in terms of numbers of organisations, significant losses might occur in terms of mobilisation of community action, grass-roots opinion and the democratic basis of the organisational structure, if these local organisations were to disappear through natural wastage or amalgamation. Nonetheless, all three countries constitute a hierarchical system capable of delivering grass-roots opinion to the central decisionmaking institution although in both Spain and the UK the final link in the chain of information and advice is missing. Likewise organisations with responsibilities for a limited range of devolved management functions, mainly in the field of first-hand sales, occur in all the case countries.

Hierarchical Governing

Intervention systems are the most classical and characteristic mode of governing interaction between the state and its citizens, as individuals or in organised form. The most common and widely practised form is by way of policies. For almost any broader subject on almost every level of public involvement in social affairs, policy is standard practice. Often this involvement is closely linked with one or more forms of legal or administrative rules. Still they can be distinguished as separate forms of interventionist governing.

Policies aim at intervening in social-political situations in a rational manner on the basis of knowledge of causal relations, but at the same time are part of these situations. This rationality element is expressed in approaches such as process-stage models. Although there are many different ways in which policies have been defined as processes, stages are distinguished for each of those in some way or other: a preparation stage that is mainly concerned with constructing a problem into a policy issue; an execution stage at which alternatives are considered and choices are being made; and a consideration stage testing and executing the selected policy.

The Common Fisheries Policy is the most decisive policy in the European fisheries sector. The Policy itself is a hierarchical instrument, as part of the Community legislative system, which by itself is a hierarchical system (see Holden, 1994).

De jure and *de facto*, the state is bound as for the selection, form and application of the means it wants to use. The constitutional state requires the relations between State and citizens to be dominated by legal principles and rules. The deeper the State penetrates the private sphere, the more formal its demands and guarantees. Consequently, design and applications of interventions are also based on principles such as equality before the law, legal security and integrity of the law. Interventions by the state are subject to political scrutiny, which means consultation of social partners, which usually presents limitations to the scope of potential legal interventions. There is also the growing interconnection of interventions by law on different levels of sub-national, national and supra/national regulation.

The grip of the State on social conditions seems to have become (almost) a handicap. As intervention by legal means was extended, its limitations and disadvantages became visible. This is one of the reasons why legal intervention is coming under scrutiny - deregulation is one of the outcomes.

Many doubt the capacity of government to regulate modern, complex society. Factors supposed to cause the deficits and the failures of public policy: lack of knowledge, time and financial resources within the public service, especially enforcing agencies; the power of social groups, especially business to influence public decision making or to resist government pressure, and the complexity of the issues at stake, which complicates or even obviates the design of adequate regulatory devices. Likewise, in empirical studies on the deficits of fisheries management there are many references to the complexity of rules and regulations concerning aspects as the kind of gears to be used, maze widths, motor capacities, and periods and areas in which fishing is allowed.

Notwithstanding the fact that many regulatory efforts by public authorities are far from effective, it remains of vital importance that those authorities – at whatever level they may operate - are in the position and use this position to make rules and regulations in the classical sense of hierarchical governing in situations where this is deemed necessary. It is a central and crucial element of modern liberal-democratic thought and practice, that the state has unique powers to intervene in situations where no other social-political institution is capable of answering a certain governing need, including fisheries.

Although there may be many misgivings about the effectiveness of many of the measures taken at the different levels and the legitimacy of those measures, it is beyond any doubt that the state of the resource calls for urgent public intervention. Part of this may take the form of co-arrangements, but there will be – in the background - the availability of public intervention if other modes of governance fail. Experiences with co-management schemes in the Netherlands, as reported in the Chain case, are indications here. The institutional form of the Biesheuvel Groups could and can only be chosen with the potential threat of hierarchical intervention when push comes to shove. This attitude is based on a general political will: a large majority in Dutch Parliament deemed a real excess of the allotted quota to the Netherlands politically unacceptable. Authoritative hierarchical governing cannot be dispensed with as a potential and actual mode of governing in a (crisis) situation.

Mixed-Mode Governing

Collective problem-solving and collective opportunity-creation in complex, dynamic and diverse situations are public as well as private, governmental as well as market challenges, a set of profit as well as non-profit activities. Different actors take the lead in different situations. A growing number of social-political challenges call for shared responsibilities and 'co-arrangements'. A (preferably) thorough and combined public and private insight in the diversity, dynamics and complexity of social-political questions and the conditions under which these questions arise is indispensable to solving social-political problems and creating social-political opportunities. To state this in (basic) terms for fisheries: it is the private (market) part of the sector which carries the responsibility for the governing of the primary interactions taking place in and around the harvesting, processing, buying and selling of fish and fish products. It is up to private, non-market parties, as non-profit organisations in and around fisheries (such as fishermen's organisations, producer and consumer organisations, environmental and scientific institutions) to take care of the governing of the more organised interactions which accompany these primary processes. It is the responsibility of public organisations to see to it that problems and opportunities within and around the primary and secondary processes and structures of fisheries take place according to principles and rules that reflect common and broader system-wide interests connected with these processes.

All this is expressed in different mixes of public, public-private and private forms of interactions, organised in the three modes. Little is known about their qualities or dis-qualities. On the basis of analyses so far the mix of self- and co-governing has hardly been explored; the mix of co- and hierarchical governing as well as the mix between self and hierarchical governing have been explored in somewhat greater depth. Yet, an analytical effort on the mix of all three is still at an exploratory stage.

Conclusion

The differentiation between and integration of modes of governance is one of the fundamental issues at stake in social-political governance. Mixes of self-, co- and hierarchical forms of governing are the most general (political) decisions to be made in order to create the most effective and legitimate governance conditions. The analysis shows that a fundamental reconsideration of the pros and cons of these mixes in (European) fisheries is not a pointless exercise at all. Creating opportunities might be the first beneficiary of such an endeavour. This can be expressed as follows:

➢ *Fishermen's Organisations play a key role as linking pins in the promotion of self-governing capacities in (European) fisheries, such as in opportunity-creation within the chain as a whole;*

➢ *The current debate on co-management in fisheries which its emphasis on co-arrangements in controlling over-fishing needs to be broadened to co-governing, including co-responsibilities for opportunity-creation;*

➢ *In particular, the hierarchical way of governing has to focus at creating clear but flexible institutional conditions for mixes of the three different modes by themselves.*

Conclusion: Some Meta-governing Observations

We have now produced several building-blocks which with to analyse social-political governing of fisheries, but a building has not yet been fabricated. This we hope to do by developing 'mortar' between the building blocks to keep the whole construction together. This mortar consists of two elements: political and managerial norms and criteria. In other words, a more norm-oriented framework needs to be developed that 'binds' what the analysis has 'taken apart'.

The lack of normative elements is, of course, not completely true. At all kinds of places and moments normative preconceptions have crept into the analysis, explicitly but certainly implicitly. This can hardly can be avoided in the treatment of a subject such as social-political governing, which is drenched with values from top to bottom, and from inside to outside. In fact the concept of 'social-political governing' as such is much more than just an analytical concept. It is a highly normatively charged concept and the whole development of the conceptualisation until today can be seen, if you will, as a normative exercise: 'governing' as such is out there; it is up to us to phrase it in terms of social-political interactions. The underlying normative presumption, then, is twofold: first that social-political governing is a sensible thing to do in situations and contexts which are basically diverse, dynamic and complex; and secondly that this kind of governing is not 'professional' or 'formal' thing governors do, but something which can and should be seen as a 'systems quality', a quality of the total patterns of interactions taking place between those governing and those governed.

Translating this into the governance of fisheries in Europe, the quality of its governance consists of basically two components:
➢ because of its diversity, dynamics and complexity, a social-political approach to fisheries governance is a must;
➢ the social-political character of this governance means that those governing fisheries as well as those governed in fisheries – be they public or private parties, together are responsible for the quality of governing fisheries.

This governance quality is best phrased as 'governability'. The governability of fisheries expresses not only how it governs itself (as a whole) but also how it wants to govern itself (as a whole). The empirical and the normative must be matched somewhere, and the concept of governability is where it can be done most explicitly.

Governability, then, can be seen as an expression of the way the different elements such as interactions (images, instruments and action), orders (1st and 2nd), and modes (self-, co- and hierarchical governing) contribute to governance, each in its own right but especially in conjunction. So how do you judge and evaluate these contributions? This is where the two sets of criteria come into the picture, the political and the managerial: how legitimate is this governance and the elements it consists of? And how effective is it in the face of problem-solving, opportunity-creation and the institutional conditions under which these processes are taking place? In applying criteria like this, we should not be too rational: not all problems can be solved, situations in which to create opportunities may be rare; not all governing actions can be equally legitimate, and not all measures taken equally effective. However, the issues themselves and the dilemmas they tend to involve have to be subject to governing: meta-governing.

However highbrow and far-fetched all this may sound, in the reality of fisheries governance these issues pop up every day, they are quite regularly debated and sometimes even subject to decision making. However the social-political governance approach claims that these governability questions should also be treated systematically, and efforts to answer them be given in terms of political and managerial criteria. I do not pretend that this is easy; the foregoing analysis will have shown, however, that in and for European fisheries today and tomorrow, including the re-drafting of a new Common Fisheries Policy, 'meta' considerations have to be made explicit. This is not an impossible task. We have found many enlightening experiences of the way fisheries in Europe are governed - and they can be multiplied by many others, as can the scholarly and practical insights presented in this book. Good governance as defined here can be expressed as a willingness to learn from experiences and a willingness to act on the basis of these insights. The following Chapters will discuss opportunities for learning and action.

PART III
CROSSING BOUNDARIES

Crossing Boundaries
Introduction to Part III

This third part of the book continues the discussion of opportunities, but now in a more explorative sense. The first study, by Jacqueline McGlade (Chapter 8), addresses the role of scientific advice. Advice on quota-setting is one the major pillars of current fisheries policy and as such an important (public) governing instrument. In this contribution, she sketches the dilemmas with which this scientific advice is surrounded. She fails to see why straightforward solutions should not exist, but feels that much more intensive interactions between all those involved in fisheries are needed. Modelling may be a means to make the outcome of those interactions understandable to those participating and as such it may create opportunities for dialogue instead of continuing monologues.

The second exploration (Chapter 9) stresses the need for institutional design of fisheries management. Jean-Paul Troadec sketches the outlines of a new institutional framework for fisheries management based upon the diversity, dynamics and complexity of the resource itself. His essay should be looked at as the outline of a whole set of new ideas on 'resource-based' management of fisheries and the associated institutional 'design parameters'.

In Chapter 10, Martijn van Vliet and Peter Friis explore the opportunities for fishermen to engage in chain-wide strategies. The past few decades have witnessed changes in the European fish chain that are led by the retail sector, the most innovative force in the food chain. An increasing involvement in post-harvest developments could be beneficial to the whole fish chain including fishermen themselves.

In Chapter 11, Juan-Luis Suarez de Vivero discusses the decentralisation and devolution of fisheries management authority from the central to the regional administrative level. The problem of institutions is that once they have been in existence for some time, they are taken as given. The status quo will always have its defenders, who see it to be in their interests to maintain the established institutional order. There is also an obvious risk that decentralisation and devolution of management responsibility to lower levels of administration may cause disorder, fragmentation and lack of co-ordination. No devolved system of management can avoid addressing these problems.

8 Bridging Disciplines
The Role of Scientific Advice, Especially Biological Modelling

JACQUELINE MCGLADE

There is now growing evidence from fisheries around the world that the current scientific methods and associated data used to provide advice for fish-stock management are failing systematically. The main reason is that the underlying models do not address some of the most important problems in fisheries: the complex dynamics of marine ecosystems, the unreliable nature of prediction in any spatially extended dynamical system, the propagation of errors and uncertainties in the models as a result of imperfect information about the system and its functioning, and the long and short-term risks of overexploitation associated with over-investment, cultural dependence and technical innovation in fisheries.

The most widely used models include virtual population analysis (VPA), Gulland's sequential population analysis (SPA), cohort analysis (CA) and its derivatives, multiplicative models, multispecies cohort and virtual population analysis methods (MSVPA), Schaefer's surplus production used to calculate the maximum sustainable yield (MSY) and Beverton and Holt's yield per recruit calculation (see McGlade and Shepherd, 1992). The different models have their own strengths and weaknesses, depending on the parameters they aim to estimate and the techniques used to solve them.

The main data inputs are the catch and length at age of a population or stock, aggregated in some form e.g. by area or fishery. A distinction occurs between *landings*, i.e. fish actually taken into port and recorded, *catch*, i.e. the fish actually taken from the water but not necessarily landed or declared and the *population or stock data*, i.e. fish sampled from research vessels using gear designed to estimate abundance of all the year-classes present. The basic sampling stratification of fish in the ports depends on the species, gear, origin of catch and time of year - the more complex the fishery, the more samples are required. However, the system of port-sampling relies heavily upon the veracity of the reported origin and size of landings. Therefore as by-catch and quota allocation infringements increase with increasing economic pressures, the use of primary data sets at finer resolutions cannot be assumed to produce reliable results.

Other data needed for existing assessment models include:
➢ the annual commercial effort for each gear and or vessel combination;
➢ population biomass estimates from research trawl or acoustic surveys;
➢ biological parameters such as age-specific weight and fecundity;
➢ stomach contents of commercially important species;
➢ young of the year and larval survey data to estimate year class strength and recruitment to the fishery.

There is also an implicit assumption that the stock boundaries are known and that emigration and immigration are not at issue. All of the data listed are generally sparse and insufficiently resolved to provide detailed insights into the gear-species interactions, ecosystem dynamics and short-term fluctuations in the abundance of different species. And whilst the genetic basis for some commercially important species have been established, estimates of the movement of individuals within metapopulations, the identification of stocks potentially at risk of extinction and the fine-scale genetic structures of other species using molecular techniques have not been widely investigated (Stokes *et al.*, 1993).

Current models do not explicitly model the fish birth, but instead use either a long term estimate of recruitment for the age at which members of a cohort usually appear in a fishery, or data on young-of-the-year collected separately. There are four widely-used stock-recruitment functions: a simple power function, Ricker's function, Beverton and Holt's function and the Shepherd function; however, it can generally be concluded that current data on the dynamics of pre-recruited fish is insufficient to properly parameterise these stock-recruitment models. Recent research suggests that recruitment is related to overall stock abundance, and that although depensation is sometimes observed (i.e. that *per capita* reproductive success decreases at low population levels), there is still some debate as to whether the effects of overfishing are reversible (Myers *et al.*, 1995; Cook *et al.*, 1997).

The convention for death is to use the sum of natural mortality and fishing mortality. Estimates of natural mortality have been the focus of concern in ICES (the International Council for the Exploration of the Sea) during recent years, as the potential effects of predation on young fishes have become more widely recognised. But, in general the level of natural mortality is kept constant for the fully-recruited population; a separability assumption is then made, which allows the differentiation of fishing mortality into age-specific and year-specific components. The age-specific component has many names: *availability, partial recruitment, exploitation pattern, relative level of fishing vulnerability*, or *selectivity coefficient*. The year-specific component is known as the *effective effort multiplier, fully recruited fishing mortality* or *fully exploited fishing mortality*. Obviously, the ecosystem is a highly dynamic entity, and so using long-term averages for recruitment and aggregated mortality estimates reduces the ability to understand the effects of fluctuations in other parts of the system (e.g. changes in predator and fishing behaviour) on the stability of the model.

Finally, survival is modelled according to early formulations of exponential survival using a discrete-time version of development because fish generations are considered to be separate; the ageing process is thus taken in discrete steps of one year. This approach removes the need for an explicit growth model, by instead using an average weight-at-age vector.

The derivation of sequential population analysis (SPA) in 1965 by Gulland, and its simpler form in 1975, cohort analysis (CA), in a sense transformed the procedures to calculate the yields of fish that could be generated from a fishery. Based on the catch equation of Baranov, and using reasonable estimates of fishing and natural mortality to begin an iterative procedure, SPA allowed scientists to get an estimate of the fishing mortality, independent of effort data, which best fitted the observed catches given in numbers at age in a year class. For any cohort, the SPA model expressed the ratio of population abundance to catch as a non-linear function of fishing mortality. The important point was that the relationship between catch in one year and numbers in the next referenced the same point in time. But there were many problems with the SPA model: it had too many parameters, the solution procedure did not produce unique parameter values, large errors occurred in the estimates of the numbers and fishing mortality at age, generated by fluctuations in the natural mortality, and the results were highly sensitive to the terminal fishing mortality - which was a guess, anyway.

The less complicated cohort analysis model solved some of these problems by introducing a discrete approximation into the continuous exponential survival model so that the entire catch is taken exactly midway through the year. With this approximation it was possible to calculate population abundance estimates and vital population rates directly from the catch data without any need for an iterative procedure, a computer or reference to virtual populations. However, there was no objective function for cohort analysis, and because the method was entirely deterministic, there was no way of telling how well the parameters were estimated. Nevertheless, the simplicity of this model made it widely accessible to many practising fisheries biologists.

The problem of too many parameters was supposedly changed by the introduction of a separability assumption. The idea behind the separability assumption was that in any year fishing mortality could be described by two factors, a full-recruitment fishing mortality or exploitation pattern and a factor to account for the differential effect of the annual exploitation pattern on various age groups in the stock. The separability assumption has become almost a standard feature of newer models.

Conventional and separable VPA models perform reasonably well at estimating stock size and fishing mortality, and thus form the basis for short-term catch predictions. However this must not be taken naively to mean that there is no need to collect auxiliary data such as effort or catch per unit effort. Indeed there are many difficulties associated with trying to simultaneously estimate stock size and fishing mortality from catch-at-age data, which contain not only errors of collection but imprecision due to collusion by industry.

A range of methods have been developed to calibrate the results from stock assessment models with effort data or some other fishery independent data. This approach is often referred to as *tuning* of virtual population analyses. But all these methods are *ad hoc* and do not assume any explicit relationship or mechanism between effort and fishing mortality, thus limiting their use as predictive tools.

The models above are aimed at single species. Anderson and Ursin provided the theoretical foundation for a multispecies approach by trying to incorporate simple predator prey interactions into fisheries management. At first, the reaction of ICES was one of reserved interest until the papers by Ursin eventually brought the concept of variable predator driven mortality into the working groups and into ICES at its annual meeting in 1979. The main aim of the multispecies approach was to apportion natural mortalities into those derived from predation by fish species of known biomass (derived from stomach analyses), and those derived from other causes.

Despite the real insights to be gained from running a multispecies virtual population analysis, little real effort has been put into improving the level of data collected for implementing MSVPA and other ecosystem models such as ECOPATH (Christensen and Pauly, 1992; Christensen, 1992). Thus the main problems with MSVPA still exist: many of the simulation runs contain a number of inaccuracies caused by a shortage of data and the need to repeat the use of data over many years.

How Well Do the Models Perform Overall?

In today's institutions there is a belief that the effects of intervention in fisheries can be predicted and hence performance measured, but unfortunately, as hindcasts have consistently shown, most current information about fish stocks is very inaccurate.

The supposition arises because the current, largely deterministic models have been set up in such a way as to allow managers to simulate and therefore anticipate future scenarios for particular real-world situations. Implicit is the idea that the models can successfully imitate not only the observed changes in the system, but also the processes that will direct its forward evolution. This supposes that the elements within a system do not change their behaviour, and that any future states are already contained and somehow defined in the present system. This is clearly unrealistic. First the inner dimensions of such a model would have to contain so much working detail that in practice it could not be developed, and second the outer dimensions could not correspond with the fact that complex living systems are open and hence experience exchanges across their boundaries.

Sensitivity analyses of the effects of systematic errors reveal that the length of the time series available or the historical level of fishing mortality limits the usefulness of cohort analysis for the estimation of stock size in the most recent years. Recruitment and biomass estimates are sensitive to errors in the independent index of stock size used for the calibration of cohort analysis. Errors affecting the age composition of the catch can lead to an overestimation of recruitment in years when recruitment is low, but to an underestimation in years when it is high. Because of this, recruitment estimates from cohort analysis tend to be less than the true quantities. Multispecies models can be developed quite easily, but the necessary data are not available in most regions: moreover, the impact of top marine predators such as seals and dolphins are not included. Where there are significant species interactions single-species yield curves are very likely to be misleading and should not be used for generating advice. The data sets used in assessments contain insufficient information to really undertake a thorough analysis of the underlying spatial and temporal changes in the real ecosystem. Potentially worse is the conclusion of Brander (MAFF, Lowestoft) who showed that despite the increase in data over time there was no indication that the forecasts were getting any more or less precise.

To date, technical controls have not managed to constrain or restrict capacity or exploitation. The reasons are manifold, reflecting the fact that fisheries are part of a highly complex, spatially extended system about which we know very little. It must also be remembered that the outputs from these models lead directly into technical instruments (e.g. Total Allowable Catches and gear restrictions); the fact that they are generally analysed in a policy vacuum means that they can not only impede the use of other more effective measures for managing stocks but also lead to highly uncertain outcomes, as in distortions to the inherent dynamics of the system (Allen and McGlade, 1987) or collapse of a stock (Hutchings *et al.*, 1997a).

The major problem is that all the current assessment models place strong emphasis on the biological rather than human or economic aspects of fisheries (Allen and McGlade, 1987). This one-sided approach is even more unbalanced when we recognise that the biological data merely refer to commercial fish species, ignoring the other components of the ecosystem, or any long-term effects associated with environmental change, pollution or human activities.

The approach is mainly based on historical grounds, which has been underscored by an increasing interest in the details of the models themselves and their statistical rigour rather than a revision in approach to incorporate new ideas and understanding about why fish stocks fluctuate and the impacts that different local, regional and global processes e.g. global trade, climate change, basin-wide oceanographic dynamics and pollution, might have on catches and recruitment. One of the implications is that the biological constants which are a fundamental part of the scientific models undergo a process of remeasurement and redefinition, sometimes gaining precision but often leaping about to an alarming degree. Improved understanding and design of experiments, new instruments and theories all combine to produce an erratic path of values, which become embedded in the literature at different times. Scientists try to cope with this by using error bars. But because the errors derive from uncertainties due to sparse data collection, equipment calibration, sampling design, and a lack of skill and confusion about the theoretical foundations of the causalities embedded in the model, the behaviour of the models is difficult to interpret.

If the situation is highly complex, as in many fisheries, simple inexactness does not fully describe the problem, for uncertainty is not merely the spread of data around a mean we know with confidence. Rather a systemic error, rendering results unreliable, may swamp an easily calculated random component in the model. So that even in what appears to be a quantitative science, achieving certainty relies largely on managing the different sorts of uncertainty affecting a large number of processes.

What Can Be Done to Improve Fisheries Management Advice?

Three scientific aspects need to be addressed in the current management process:
➢ the extent to which uncertainty and risk might invalidate the models underlying fisheries policies;
➢ an assessment of the difficulty of the problems relating to fisheries;
➢ the methods used to adjudge the performance and effectiveness of different fishing practices and policies.

In fisheries, the objectives of conservation and management imply altering fishing mortality or recruitment to the fishery through effort regulation and gear selection. The target values of these objectives have traditionally been derived from deterministic models such as simple yield per recruit calculations, stock-recruitment and density-dependent relationships, the choice depending largely on the type of data available. But of the 100+ stocks for which the EU sets Total Allowable Catches (TACs), only 40% are based upon assessments which permit catch forecasts to be calculated for the coming year. The others are based upon either long-term average catches or average historic catches and are termed "precautionary TACs". Taken at face value this suggests that there is still a high degree of uncertainty in the advice given, especially when put into the context of the political trading environment amongst the European Council of Ministers.

When quantitative sciences are used to provide input for the policy process, as in fisheries, the problems of managing uncertainty are severe. First, the original data are not those of a controlled laboratory environment. Well-structured theories, common in many branches of science, are conspicuous by their absence. Furthermore, fisheries research is really an interdisciplinary activity, involving fields of varying states of maturity and with very different practices. Scientists need to use inputs from a variety of fields they are likely to be unfamiliar with; they thus find it difficult to make the same sensitive quality judgements as in their own field. The result is a dilution of quality control on the research process and a weaker quality assurance of results for setting fisheries policy.

This situation is compounded by the fact that scientists rarely receive training in quality assessment; they therefore develop a healthy prudence about passing judgement on the results of others and any tendency to meddle in others' fields is discouraged. Unfortunately, in an interdisciplinary policy-related area of research, such tact and reticence can be very counterproductive, because criticism (the lifeblood of science) does not occur in sufficient strength. What is needed are new ways to evaluate scientific work to provide clear, explicit and public guidelines for analysis and communication, while retaining the skills and judgements of the best traditional sciences.

Their public dimension increases the problems of uncertainty in policy-related research. Fisheries science is judged by the public, including bureaucrats, on its performance in such sensitive areas as restrictions on fishing gears, operational practices and by-catches of marine mammals. All these involve much uncertainty, as well as inescapable social and ethical aspects. Simplicity and precision in predictions and setting limits are not feasible, yet policy-makers tend to expect straight-forward information as input to their own decision-making process.

In such circumstances, the maintenance of confidence among policy-makers and the general public becomes increasingly strained, with the scientist often caught in the middle. The problem becomes manifest at several levels, the simplest one being the representation of uncertainty in quantitative estimates. The scientific adviser knows that a prediction such as a 'one in a million' chance of a serious event such as the collapse of a fishery should be hedged with statements of many sorts of uncertainty, so as to caution any user on the reliability limits of the numerical assertions. If these are all expressed in prose, the statements become tedious and incomprehensible to the user, but if they are omitted or even given in some simple statistical representation, the same adviser can be accused of conveying an unsubstantiated certainty (Hutchings *et al.*, 1997a, b; Healey, 1997).

The same dilemma occurs whenever scientists give advice on policy-related issues: any definite advice is liable to go wrong, a prediction of danger will appear alarmist if nothing happens in the short term, and reassurance can be condemned if it retrospectively turns out to be wrong. If the scientist prudently refuses to accept vague or even qualitative expert opinions as a basis for quantitative assessments, and declines to provide definitive advice when asked, then science is regarded as obstructionist and not performing its public functions. Thus the credibility of science, based for so long on the supposed certainty of its conclusions, is endangered by giving scientific advice on such inherently uncertain issues.

Is There a Way Out of This Dilemma?

Uncertainty cannot be removed, so it must be clarified. The errors associated with data points, usually expressed as error bars and significant digits represent a spread, i.e. a tolerance or a random error in a calculated measurement - a measure of inexactness. Another type of uncertainty relates to confidence that scientists have about an observation or linkage; thus safety and reliability estimates are given as conservative by a factor. In risk analyses and future scenarios estimates are qualified as optimistic or pessimistic. In laboratory practice the systematic error in physical quantities as distinct from the random error or spread, is estimated on an historical basis. Thus it provides an assessment to act as a qualifier on the number and also on the spread. We can use techniques such as fuzzy-logic to characterise these types of uncertainties and also the level of ignorance about the system.

Ignorance is the measure of the gaps in our knowledge; these may simply be anomalous results that are exposed when a new advance in understanding occurs. The boundary of ignorance is very difficult to map, and relies on the pedigree of a particular field, i.e. the degree of consensus amongst individuals about the paradigms of the discipline and whether its results are assessed statistically from historical data; based on statistics as far as possible or simply guessed. Fisheries probably fall between the first two: making predictions under such circumstances is thus rather premature and gives a misdirected sense of concreteness in policy judgement. Any certainty or precision about the status of different fish stocks must now compete with the flexibility and ambiguity needed to present scientific advice. The virtue of simplicity and the sinecure of the single prediction have to give way to the modern need to reflect complexity and uncertainty in technological development and resource management.

In policy-related research, the credibility of science is at risk, because of the dilemmas of uncertainty and responsibility that confront scientists. These cannot be eliminated, nor indeed would that be desirable. Managing uncertainty in the context of responsibility cannot be side-stepped, but unfortunately, the institutions we have developed to control these problems have been designed from a perspective of determinacy. Institutional behaviour is also biased away from fishermen, who have not wanted to be directly involved in management unless a resource is in danger of collapsing. As such a valuable source of data is generally lost or compromised.

A new world-wide development aimed at addressing many of the problems described above is the soft intelligence system called SimCoast (McGlade and Hogarth, 1997). Inputs include data from resource surveys, environmental monitoring, ecosystems analyses, dynamical models, global time series, socio-economic and policy analyses. Through directed workshops in South-east Asia, Africa, North America and Europe, experts involved in different arenas of resource use and management, including the commercial and industrial sectors, come together and agree on the conditions under which fisheries and other resource activities are being governed. Rules, relying where necessary on fuzzy logic, are derived and used to examine the consequences of different policy and management scenarios. The process is highly inclusive and creates a level playing field between disciplines and participants; in this way information from the fishing industry can be set alongside model outputs and evaluated in a transparent manner through examination of the expert rules and the certainty factors attached to them. Moreover, the global linkage means that fleet dynamics and the impacts of world trade can also be built explicitly into different scenarios and policies.

The use of such wide-input approaches to management is important because existing fisheries policies in Europe do not specifically include performance objectives other than keeping catches within specified levels or stocks above a minimum spawning stock size. For example, there are no objectives or incentives to decrease the amount of damage done by fishing gear such as beam trawls on benthic communities or to evaluate the long-term health and genetic effects of fishing on different fish stocks. Although the idea of a Precautionary Principle is essentially part of fisheries policy, the approaches used do not take into account the fact that continued exploitation at the levels of fishing we are witnessing now are anything but precautionary. TACs might in principle be precautionary, but cannot be so in reality when they are being implemented in the current climate of overcapacity. The words and actions do not match.

One crucial element in the formation of conservation policy and its associated economic structuring is the evaluation of events occurring at different points in time, the most obvious example being the use of discount factors in valuation or optimisation. In ecological terms, the longer a policy on conservation takes to come into effect, the more deleterious the state of the resource or ecosystem might become if it is already in some danger of over-exploitation. The risk then lies in the future pay-offs and how quickly they are likely to decline or accrue. If, as we have discussed above, the fishing industry is simply part of an evolving dynamical system, then it is not clear that the processes at work will always present the same biological cocktail of species. In fact it is quite likely that the unending exploitation of certain segments of the ecosystem will lead to change, and to some extent degradation. Thus instead of an increasing positive discount function, typical of most economic analyses, delaying decisions about conservation may in fact give rise to a negative discount function. Add to this the uncertainties of environmental degradation, and it is clear that future analysis of fisheries must acknowledge that there are stochastic processes at work in the system each with a time-dependent probability distribution. Thus if we consider the dynamics of any fisheries system, standard economic utility functions are going to lead to completely erroneous management plans.

Thus in thinking about which models and approaches to use in fisheries management, we should recall three propositions: i) not everything is possible; ii) not everything that is possible is desirable and iii) not everything that is both possible and desirable is socially or economically viable. However, criteria for ruling out alternatives have not been well-studied. It seems that whatever ecologists say is impossible or philosophers say is undesirable, economists consider to be uneconomic! This position might be unassailable if it weren't for the fact that the monetary measures of opportunity costs and information about competition, utility functions and production functions are neither accurately known nor complete in fisheries. Indeed, the idea of a rational manager making policy decisions armed with a complete set of data could hardly be further from reality. Instead we see delayed rationality at work, putting off decisions, especially if they have large economic or political implications, for as long as possible. The only way out of the loop is to create linkages using systems such as SimCoast to create a proper dialogue between the researchers, industry, politicians and society to determine which scenarios are acceptable and viable.

9 Institutionalising Ecological Boundaries

A Perspective from France

JEAN-PAUL TROADEC

Introduction

Throughout history, the scope of fisheries management has been subject to marked changes. In pre-agricultural, pre-merchant societies, resource conservation was not a critical issue (Sahlins, 1974). The defense of hunting and fishing territories, the distribution of harvesting sites among community members, or uncertainty reduction in family harvests were greater concerns. During the expansion of commercial fisheries when extensification provided an easy solution to overcome local gluts, fishery management focused on the preservation of stock productivity by adjusting the fishing pattern through technical measures - protection of females, gear selectivity, closed seasons and areas, and so on (Garcia, 1989). Now that a majority of fish stocks are fully or over-exploited on a global scale (FAO, 1992; Garcia and Newton, 1996), adjusting the fishing capacities and effort to the potential yield of fish stocks seems to be the main challenge (Troadec, 1994; 1996).

This is to overlook the importance of building up an effective capacity to regulate access. Under the new conditions of resource scarcity, stock conservation, economic rationalisation of exploitation and prevention of conflicts depend on its successful implementation. The problem is not restricted to fisheries. Many uses of renewable marine resources are confronted by the same issue: shellfish culture confronted by the limited carrying capacity of basins (Héral et al., 1989), or the disposal of human waste exceeding the assimilating capacity of the environment.

Technical measures aiming at sustaining fish stock productivity or ecosystem carrying capacity are insufficient to regulate access. Once resources can no longer meet the demand arising from their various uses, special institutional innovations are needed that match the ecological features of marine living resources, the technical characteristics of their uses, and the economic and social organisation of their users. On the other hand, since the motivation for malpractices – for example, catching under-sized fish -

187

declines in the absence of over-capacity, access regulation can significantly ease the enforcement of technical measures.

The following Section reviews some of the implications that the features of marine renewable resources have on the design of institutions which condition the regulation of access.

The Dynamics of Over-fishing and the Governing of Access

The problems of degrading fish stocks, economic wastes resulting from excessive fishing capacities, and conflicts among fishermen's groups confronted to stocks lower than their demand have a single origin: the shortcomings of the management institutions and methods developed when fishery resources were not binding and overall harvesting did not require to be adjusted to stock potential yields (Fig. 1 – all Figures: see end of Chapter). Institutions adapted to access regulation need to provide effective solutions to a double problem (Fig. 3):

First, the disparity between the space and time scales of fish stocks, on the one hand, and those of the fishing units on the other hand. Fish stocks have geographical scales that, in most cases, exceed those of the operations of fishing companies and small-scale fishermen's groups by far. This disparity of scale prevents the solution that agriculture adopted for crossing the limit of extensification - the allotment of the natural resource – being transposed to the realm of fisheries.

A similar disparity exists with respect to time. In general, fishing companies and small-scale fishermen's groups do not have the capacity to adjust to the fluctuations in abundance of individual stocks resulting from environmentally-induced changes in their recruitment. They overcome this constraint by shifting target species through changes in fishing gear and geographical displacements. Aquaculture has overcome this constraint by the control of stock reproduction.

Second, the disparity between the mechanisms used for allocating fishing capacities, i.e. human inputs, on the one hand, and the resource - i.e. the natural factor of production - on the other hand. Schematically, in fisheries and other commercial uses of renewable marine resources - including small-scale fisheries in developing countries, boats, crews and consumable goods are mobilised by fishing units through mechanisms that are increasingly and predominantly economic in nature. On the other hand, access is traditionally regulated at the scale of resource units by national authorities through command mechanisms (norms and decisions).

Under these conditions, neither harvesters, nor administrations are in a position to adjust access to the levels corresponding to the objective of economic efficiency or the imperative of conservation. The command mode does not integrate, nor rely upon, the economic forces at the origin of overexploitation. In the absence of an adequate property regime, the resource rent, which expresses the value that the resource takes in becoming scarce, is not distinguished from the normal profit that users derive from the application and management of the human factors of production. Because the resource rent accrues in addition to the normal profit of harvesters, this surplus profit stimulates them to acquire additional capacity. The overcapacities generated by this behaviour leads, first, to the dissipation of the resource rent, then to overfishing (Fig. 2).

The efficacy of the command mode is low in commercial activities where decisions have direct and explicit distributive effects. This is the case of access regulation. In theory, technical measures aiming to sustain the productivity or capacity of natural resources are equally applicable to all users. In practice, some have distributive effects[1], but those are generally indirect and less marked. In the absence of overcapacity, they can, therefore, be less difficult to enforce by command. The great difficulties that management authorities encounter in achieving the objectives of economic rationalisation, resource conservation and conflict reduction appearing in national fisheries policies, even in countries with well-established scientific and administrative capabilities, is consonant with this interpretation.

On the other hand, because of fish stock mobility, individual fishermen cannot adjust their fishing capacities to the capacity of the part of the stocks they exploit. They know that such behaviour would lead to the attrition of their share. To be effective, such an adjustment should be made on the scale of the stocks - i.e. either by the collective of fishing units participating in the exploitation of a stock, or by the public sector. The applicability of the first option, however, is declining with the dislocation of traditional fishing communities, the growing professional (in- and out-) mobility and the diversification of uses.

Thus, a double system of exclusivity rights is required for regulating access in the exploitation of common renewable resources. Individual user rights are quantitatively and qualitatively defined; if, in the great majority of cases, resource units cannot be allotted directly, use rights can be defined on inputs and/or outputs - on the control variables of the use-specific exploitation functions (fishing, extensive aquaculture, pollution and other (ab)uses of the environment). The other defines the rights of the administrative body or user group to impose ceilings on individual rights, and the mechanisms (market, open access plus fee, command) through which the limited amount of rights will be allocated among candidate harvesters.

The coincidence between the geographic competence of management structure and the spatial distribution of the resources is, thus, an important condition of management effectiveness. Two units dominate the structure of aquatic renewable resources: the ecosystem and the population[2]. However, the adjustment of regulations is complicated by interactions that blur the resource structure:

➤ *technological interactions* resulting from the low selectivity of fishing gear which does not permit the separate exploitation of fish species and populations;

➤ *biological interactions* due to which fishing one species may have repercussions on its prey, predator and competitor species;

➤ *ecological interactions* within ecosystems, which causes the impact of one use to spill over to other uses (e.g. negative and positive effects of pollution on the carrying capacity of shellfish basins).

But the operational difficulties resulting from these interactions do not detract from the basic principle. The structure of populations and ecosystems is independent of their uses and, despite significant year-to-year fluctuations, geographically stable. These are essential assets for the design of proper management structures.

As a result, the problem of overfishing is well understood (Fig. 2). Basically it is an economic problem of accounting for the natural factor in activities which are predominantly commercial. Its solution, however, depends on the prior adjustment of institutions to the new conditions of resource scarcity. The fact that, conceptually, the problem is clear, does not mean its solution is necessarily simple. Because of the differences in the ecological and biological features of the resources, in the technological characteristics of their uses, and in the economic and social organisation of users (Table 1), there is no standard solution to the regulation of access in fisheries and other uses of marine renewable resources. Furthermore, the adoption of institutional innovations is uncertain. Since institutional reforms have significative distributive effects, they face strong cultural and political opposition, even when it is clear that they will lead to significative improvements in efficiency, equity and effectiveness.

Current Exclusivity Schemes

Public Sovereignty

On land, the exploitation of natural resources is ruled by a two-pronged system of complementary rights. State sovereignty guarantees and defines individual property systems through basic state functions (army, judiciary and police). The situation at sea is not fundamentally different from this scheme. The main difference rests in the lesser advancement of sovereignty and property systems. The new Ocean Regime allows coastal states to devise and enforce legislations to regulate access to the resources that are distributed within their areas of jurisdiction they have recently acquired. In the area of fisheries, most countries have revised their legislations applicable to the activities of foreign fleets, their authority with respect to which is now clear. On a global scale with notable exceptions, and in the European Union in particular, they have been slower to adjust their legislations applicable to their domestic fleets. This is particularly true of small-scale fisheries in which rural groups have a long tradition of inhabiting littoral areas and exploitation of coastal stocks.

Despite this development in international law, national authority remains diluted from the shore to the high seas: especially in their territorial waters, coastal states have quasi-property rights over the resources that are fully circumscribed within their respective Exclusive Economic Zones (EEZs), but they share this authority for stocks whose distribution area extends outside that zone. Here two situations have to be distinguished: authority of coastal states is closed for shared resources that are fully circumscribed within their EEZs, or it is open for resources which extend beyond the EEZs into the high seas.

The historical trend towards offshore extension of national jurisdictions may not be completed yet. The case of straddling stocks remains unsettled. The extension of national sovereignty may, however, not go much beyond the continental slope within which most fisheries, as well as other resources that are currently exploited, are distributed. For the conservation and management of oceanic resources such as tuna fish and whales, the delegation of authority to supranational structures may be a better solution.

In any case, the dilution of national authority in offshore waters is not the most critical constraint. With some exceptions, coastal countries have only in part reaped the opportunities created by the Law of the Sea.

The scope of this recent development is far-reaching, for both at international and national levels, voluntary agreements adopted by groups of countries or users outside legal frameworks remain vulnerable to the arrival of newcomers. The new Ocean Regime provides the sovereignty framework within which coastal countries can, separately and jointly for shared stocks, build up the property schemes that condition progress in the management of their domestic fisheries. Governments have a particular responsibility in this respect since, in the historical development of property systems and institutions ruling the exploitation of natural resources and economic activities, public solutions have ultimately prevailed over collective initiatives. To constrain individual free-riding, public solutions have proved less expensive and more effective than collective arrangements (North and Thomas, 1973).

Property Schemes

An explicit distinction between resource property and use rights, and the allocation of exclusive property rights to the State and of temporary use rights to private - individual or collective - actors would provide an operational solution to the disparity of scales between resource and fishing units. This distinction would facilitate the extraction of resource rent and the mastering of overcapacities. It would allow to maintain the fleet-stock mobility needed to offset the consequences on the fishing performance of fluctuations in stock recruitment. It would enable integrated management of renewable resources that subject to multiple uses.

The principle associates management with resource ownership. It makes the State formally responsible for the long-term conservation and enhancement (establishment of marine reserves, eradication of non-indigeneous species, development of mariculture, and so on) of the natural inheritance, of the delimitation of management units, of the choice of preferred uses and the determination of exploitation regimes applicable to different uses of resource units (including the reservation of forage food for wildlife), of the selection and implementation of mechanisms for allocating private use rights, and of the enforcement of management schemes. It entitles user groups to lodge complaints when they believe their rights are not properly respected.

Public ownership of fishery and environmental resources does not imply that their management need to be conducted by central government. The task could be assigned to specialised autonomous authorities, as it is frequently the case in forestry, water resource management, or river basin management.

One condition of management effectiveness is that the spatial distribution of property and management structures matches the spatial distribution of resources. The distribution of fish stocks is sometimes advocated to justify the centralisation of fisheries management at the level of the European Commission and its Member States. The argument merits careful checking since, according to the principle of subsidiarity, management effectiveness is likely to dwindle as the number and diversity of parties increase, and the concern for management of local user groups, leaders and administrations declines.

Let us examine this issue by considering the situation in the English Channel and contiguous seas (North Sea, Irish Sea, and Bay of Biscay) for which the information available on the identity and distribution of major fish stocks has been recently compiled (IFREMER-MAAF, 1995). This (admittedly rough) information, summarised in Table 2, concerns stocks that can be identified under the conventional management approach, not the individual populations within each species. Improving the identification of fish populations and delimitation of their distribution areas at the different stages of their life-span is an important research topic for enhancing the discriminative power of assessment and management schemes. Yet, with the information available it is possible to appreciate to what extent the communitarisation of fisheries management is justified on ecological considerations. Indications on the geographic distribution of major uses of ecosystems (aquaculture systems, recreational activities and pollutions from land-based activities) have been added to the Table.

Coastal Resources

With the exception of river runoffs which extend further offshore, the three-mile littoral strip includes the ecosystems most directly used and affected by human activities other than fisheries (recreational activities, pollution, aquaculture). It contains unit stocks of bivalves and crustaceans; similar stocks are also found in the territorial sea; a majority of these stocks are exploited exclusively by local and national fleets of the coastal country. The shoreward extension of these stocks and ecosystems varies from a few to a few hundred miles. The concentration of fishing, aqua-cultural and recreational activities, the impacts of land-based pollution on the quality of environment and fishery and aquaculture resources, the on-going relative decline of fisheries compared to tourism and pollution, the tradition of occupation of space and the sense of ownership of littoral resources by coastal dwellers, the tradition of co-operation between local administrations and communities in coastal management, and the role played by local administrations in the regulation of land-based activities have two important consequences:

> integrated management means that fisheries should be managed (assessment, legislation and enforcement) together with the other uses of marine renewable resources;
> local authorities and user groups are directly involved in their management.

Shelf Stocks

Fish stocks distributed on the continental shelf can be subdivided in two groups:

> small stocks of restricted geographic extension (a few hundreds miles), such as cuttlefish, edible crab, sea-bass, plaice, sole and pilchard; such stocks are often exploited by a few fleets based in neighbouring countries;
> large stocks (hake, mackerel, horse mackerel, etc.) covering large areas of the Community sea; most are exploited by distant-water fleets registered in several countries of the European Union; and oceanic stocks (tuna and salmon), distributed throughout the North Atlantic and exploited by EU and third countries.

Thus, the distribution of fish stocks and other marine resources does not justify the centralisation of fisheries management at the level of the European Commission. The priority given by national administrations to employment and short-term interests of their national industries over setting up common institutions to rationalise the industry and preserve the resource, as well as the shortfall in political integration at the European level are better explanations of the difficulty of European fisheries to depart from pre-LoS conditions where national administrations could exert their authority only on domestic fleets enjoying free access to resources accross the world ocean. In this sense, the centralisation of fisheries management at the level of the European Commission has probably played against the adjustment of institutions to the new conditions.

A double set of regional management authorities, one for coastal resources and another for continental ones is compatible with the scale distribution of the resources. Such a scheme, which is consonant with the principle of subsidiarity, could facilitate the working out of solutions to two important issues:

➤ *The integrated management of multiple uses in coastal waters.* Coastal management depends primarily on the effective integration of assessment, legislation and enforcement. The development of general models simulating interactions of multiple uses of coastal ecosystems is complicated by the complexity of ecology, but models describing interactions between certain uses (e.g. shellfish culture and agriculture) are being developed. The integration of legislations applicable to different uses is not very advanced. French legislation on aquaculture, for example, contains provisions for its regulation, but those do not always bear on the primary control variables of the different production systems. If individual leases are allocated on the Maritime Public Domain, there is no provision for the regulation of stock densities in basins that are overstocked, while offal disposal in cage farming is regulated separately under the legislation on environment. The lack of coherence between existing legislations applicable to fisheries and shellfish culture impedes the development of scallop mariculture (Curtil, 1996). This shortcoming requires careful examination in the perspective of the projected abrogation in 2002 of exceptions to the common fisheries management system in the territorial sea. The establishment of integrated management authorities in coastal areas may be a solution to this issue.

➤ *The contradiction between the principle of relative stability in fisheries and the principle of free circulation of persons, commidities and services laid down in the Treaty of Rome.* The current weaknesses in property regimes do not facilitate the reconciliation of these two contradictory principles. The acceptability of the freedom of circulation principle is presently low in fisheries, since the mobility of boats it would result in is likely to lead to serious political turmoil in coastal areas. In agriculture, access to land is governed by elaborate property schemes which limit farmers' mobility and facilitate the influx of newcomers into country societies, while market mechanisms governing the allocation of land contribute to the adjustment of production capacities to its natural productivity. The establishment of a double scheme of access regulation, one for coastal (territorial sea) and one for continental areas (EEZ), could provide a solution to balance the opposite objectives of fleet mobility and social stability. Boat registration in the ports of the management area may be imposed as a precondition for eligibility for acquiring rights to fish coastal stocks. This clause may be lifted for large-scale fleets operating in continental management areas, whose mobility is high and for which the adoption of market mechanisms for allocating fishing rights may raise less objections.

Given the importance of shared stocks and common problems of environment conservation resulting from the extension of closed seas in the Community sea, the major difficulty in setting up regional management structures will be their integration into the different levels of political integration in Europe (districts or departments, regions, states, EU). The European Union could be an asset for adjusting European institutions to the new conditions. It could provide a forum in the European Commission where Member States examine the technical aspects of the issue, negotiate the establishment of a spatially coherent set of management structures, and define their mandate, membership and working rules, including the accountability of their performance.

Conclusion

Since total fish production in the Northeast Atlantic levelled off some forty years ago, resource conservation, creation of wealth, and reduction of conflicts in the fishing industry depend primarily on the regulation of access. Still, in the European Union, fisheries management remains partially based on institutions that were developed before the resources became constraining - i.e. when the application of technical measures for sustaining stock productivity became a key concern. As a consequence, effectiveness in attempts to regulate access is low.

Three sets of institutions are involved in the regulation of access: the regime of exclusive rights, the mechanisms of use-rights allocation, and the structures to implement the allocation mechanisms.

Property should not be understood as an absolute right to commodities, but as a coherent set of rights to freely and legally exert controls on the relevant variables of the use function, distributed among the different levels of public, collective and individual organisations according to their respective capacities to exert the different controls at appropriate scales. The need to clarify exclusivity systems is not specific to a particular allocation mechanism. If the sophistication and complexity of exclusivity systems increase with the specialisation and diversification of human activities that characterise modern economies, the clarification of rights and duties is also a precondition for the effectiveness of other allocation mechanisms: command, collective management, or co-management.

Fish and environment resources have features that have important implications of the design of institutions. In relation to environmental fluidity and fish mobility, the disparity between the space and time scales of fish stocks and fishing units is large, and interactions between uses through their respective impacts on ecosystems more important than on land. The consequences of these characteristics are the following:

➢ management efficacy depends on the ability to keep management institutionally separate from exploitation;

➢ this implies a distinction between resource property and use rights;

➢ the resource scale does not justify, however, the centralisation of management at the highest level of political integration; management will be more effective if effected through decentralised regional structures reflecting the resource scale;

➢ in coastal areas, integrated coastal management passes by a greater integration of systems that were developed separately for fisheries, aquaculture and environment conservation.

By clarifying sovereignty regimes, the new Ocean Regime has created conditions that countries can use to design and implement the exclusivity systems needed to enhance the effectiveness of access regulation. Compared to other countries, however, the European Union has been slow to reap this new opportunity. The defense of short-term interests of national industries by national administrations tended to prevail over the achievement of the objectives of resource conservation and economic rationalisation in the European Fisheries Policy, and the establishment of the institutions required for this purpose. In this respect, it is worth noting that island states (Australia, Iceland, New Zealand, etc.) whose fishery resources are better isolated from those in neighbouring countries, are among the leaders in the adjustment of management institutions. Still, the European Union can be an asset for the design and adoption of management institutions in an area where co-operation is imposed by the extension of closed seas and the importance of shared stocks.

Strategically, it seems advisable to start from a clarification of sovereignty and property schemes. If the approach may look politically audacious, it is technically simpler. Common exclusivity schemes could be adopted, which is not the case for use rights and allocation mechanisms whose merits differ with the resources, the uses and the user organisations under consideration.

National governments and the European Commission have a critical role to play in adjusting management institutions to the new conditions. If economic forces are a powerful factor of institutional adjustment, the delay in adopting timely institutional innovations shows that they are not sufficient. As evidenced by the changes induced by the new Ocean Regime, an agreement on sovereignty and property schemes can have cascading effects on the adjustment of management instruments and practices at lower levels of national administrations and industries.

The EU framework can be used to negotiate an integrated set of management structures whose pattern would reflect the resource distribution (littoral, territorial, shelf, etc.), and mandate the nature of priority tasks. It is, therefore, the adjustment of institutions to the new conditions, rather than the defense of the short-term interests of the industry, that the public sector should prioritise when investing its talents and energies.

Notes

1   For example, small-scale fishermen with a limited range of operations will be more affected by mesh-size regulation when smaller fishes are concentrated in coastal areas.
2   For a given species, a population represents a set of individuals that is genetically distinct from other populations within the same species and occupies a particular hydro-dynamic structure of the environment.

FIGURES AND TABLES

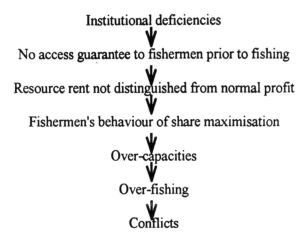

Figure 1 - The dynamics of over-fishing in commercial fisheries

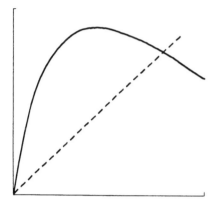

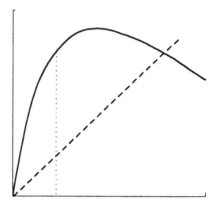

Figure 2a – Dynamics of overfishing under
open- and free access
conditions

Figure 2b – Effects of market allocation, or
of imposing a fee, on the rate
of exploitation

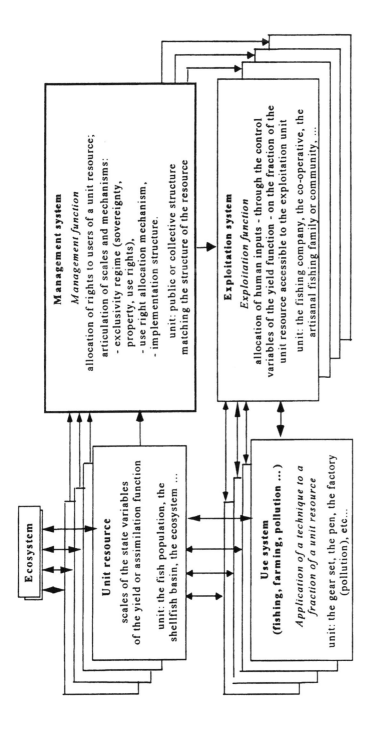

Figure 3 - Articulations between ecosystems, unit resources, and production, exploitation and management systems in uses of marine renewable resources

Table 1 - Summary of biological, technological, economic, social and political factors to consider when devising access regulation systems in fisheries. (N.B. The number of pluses and minuses in the Table is indicative. Finer distinctions between fisheries exist.)

	Stocks						Gear		Economic, social and political organisation of fishery systems		
	Mobility		Variety of Species		Recruitment				Small scale	Large scale	
Institutions	Sedenta-ry	Mobile	Mono-specific	Pluri-specific	Stable	Un-stable	Fixed	Towed		Do-mestic	Foreign
Exclusive rights (property ≠ usufruct)	+++ (1)						+++ (1)		+	++	+++
Fishing right support											
- output controls			+++	-	+++	- (2)	+++ (6)	+++	- - - (3)	++ (3)	++/- (3)
- input controls		- - -	++	++ (4)	++	++ (5)	+++		- (7)	+/- (7)	++ (7)
- territorial leases	++			+	++	- -			++ (7)	+ (7)	= (7)
Allocation mechanisms											
- administrative rationing									+/- (8)	+ (8)	++ (8)
- fixed price (fee)									++ (9)	++ (9)	++ (10)
- market										++ (11)	+++ (11)

Table 1 (ctd.)

(1) As illustrated by shellfish culture and aquaculture, sedentary species and fixed gear facilitate the allocation of territorial leases.

(2) Where scientific capacities are limited, and for stocks of lower commercial value, recruitment variability may be difficult to monitor with the desirable periodicity.

(3) Catches are more difficult to monitor than fishing capacities; this is particularly the case in small-scale fisheries, or for fleets that do not land locally.

(4) In multispecies fisheries, to limit 'high grading' and transfer of effort on unregulated species, input controls may be an alternative, or a complementary measure, to output controls.

(5) For stocks that do not show a market schooling behaviour, input controls may be an alternative, or a complementary measure, to constraint (2).

(6) Control of fishing power (gear characteristics) and effort is sometimes easier for fixed gear.

(7) As an alternative, or a complementary measure, to constraint (3).

(8) In domestic (especially small-scale) fisheries, if administrative rationing is easy to adopt, effectiveness of enforcement is limited.

(9) Taxes on exports may be a second-best solution for the part of harvests that is exported.

(10) A possible transitory solution, before a market mechanism can be adopted, also for foreign fleets which land their catches locally.

(11) In Southern countries, the clarity of property regimes with respect to foreign fleets and the capitalistic nature of the latter may make the adoption of market mechanisms (auctions) easier.

Table 2 - Tentative classification of stock and environment units based on their inshore/offshore and alongshore distribution, and its implications regarding the distribution of management duties (examples are drawn the Western part of the English Channel; IFREMER and MAAF, 1993).

Inshore/offshore distance	0 – 3 nm	3 – 12 nm	12 – 200 nm	> 200 nm
Coastal stocks	- examples of stocks: scallop and other bivalves, whelk, spider crab, lobster, etc. - alongshore extension of management areas: NW Brittany, Brittany-Normandy Bay, etc. *			
Small continental stocks		- examples of stocks: cuttlefish, edible crab, sea bass, sea bream, plaice, sole, lemon sole, pollack, gunnard, sardine, ... - alongshore extension of management areas: Western English Channel, Eastern English Channel, Irish Sea, Bay of Biscaye, ...		
Large continental stocks		- examples of stocks: hake, mackerel, horse mackerel... - extension of management areas: whole EU EEZ.		
Oceanic stocks				- example of stocks: tuna etc. - extension of management area: North Atlantic

Aquaculture	- shellfish and fish culture basins - alongshore extension of management areas (some tens of nm): river mouths and bays		
Recreational uses	- littoral ecosystems affected by, and areas used for, recreational activities - alongshore extension of management areas (10s nm): river mouths and bays		
Pollution from land and shore activities	- littoral ecosystems modified by pollution from land-based activities - alongshore extension of management areas (10s n)	- large river outflows - extension of management areas (100s nm)	

* Some of these stocks, which are distributed in semi-closed bays (e.g. certain stocks of bivalve mollusks such as scallop stocks in the Bay of Brest or St Brieuc), and have a narrower distribution (within the 0-3 nm inshore zone), are candidates for local (littoral) management.
** In Spring, cuttlefish come to spawn in inshore waters where they are exploited. The Table, which attempts to sift out dominating distribution patterns, cannot reflect certain species-specific features. The same applies to salmon, which falls in the oceanic and coastal stock categories.
NB Uninterrupted borders with text indicate main concentrations, dashed lines (dotted areas) indicate secondary concentrations.

10 Creating Co-operation in the Chain

Options for Integrating 'Catch' and 'Market'

MARTIJN VAN VLIET AND PETER FRIIS

Introduction

The food market in Western economies, in particular the European Union, is changing rapidly. Many actors involved in producing, harvesting, processing and selling food therefore feel uncertain about their future and try to re-orient themselves. The most important changes are (Grunert *et al.*, 1996: 1):

> Growth in volume is stagnating. Growth, if it occurs, is realised by adding value to the same quantity of production.
> The European Union faces excess agricultural production.
> Declining trade barriers, deregulation, and increased pressure on subsidies in the EU makes primary producers more and more vulnerable to changes in the market and increased competition.
> Increased concentration in retail and processing together with highly dynamic and quickly changing consumer behaviour have shifted the power relations towards a limited number of multinational food industries and retail chains controlling the market.

These changes in market conditions have an enormous impact on people and businesses involved in producing, transporting, processing and selling food and on the relations between these parties. It is often suggested that 'the chain is turned upside down': one has recognised that the consumer is not at the *end* of the chain but at the *beginning*. For long time agriculture and fisheries was *product-oriented*. The main question has been how to bring the products to the consumer. However, the food chain has become *demand-oriented*. More and more, the question has become what and how to produce in a way that the consumer is willing to buy the products. Such interactive information exchange between the different links of the chain have become of the utmost importance. As a result, knowledge of consumer demand and of ways to manipulate it has grown in importance. Therefore the parts of the food chain that are able to control the gateway to the consumer have gained power. Related to this tendency, we have witnessed a concentration in retail and food industry resulting in the existence of huge multinational enterprises in the food industry and retail sector. These enterprises dominate the market on the basis of their market share and knowledge of consumer behaviour.

As a result of this growing demand-orientedness, the parts of the food chain that are in or close primary production seem to have difficulty in adapting their business to the new market conditions. This does not only apply to the primary producers themselves, farmers, market gardeners and fishermen. For a long time the co-operatives, owned by the primary producers and established to process and to market agricultural products, have been the dominating forces in agriculture in countries such as France, Denmark and the Netherlands. Now, under the pressure in the food market, they feel the need to re-orient their business strategies in the face of increased international competition (Bijman *et al.*, 1997, Kristensen and Sabel, 1997).

These changes certainly have not gone unnoticed in the fish chain. Fishermen are confronted with low harvesting levels, due to administrative regulations and biological limits. As prices for substitution goods such as meat and chicken are stagnating or falling, normally higher prices do not offset lower 'production' (Friis, 1996). Trade and processing need to adapt to the higher quality standards pressed upon them by a combination of government regulations, quality standards from food and retail multinationals and consumer demand for high-quality and safe products. In the process of 'turning the food chain upside down', fishing, became more and more marginalised as a socio-economic activity, and the people involved alienated.

The post-war growth of the fish chain, based on increased volumes caught and sold, has come to an end. The sector is in need of a strategic reorientation in which the volume orientation should be abandoned. It should be replaced by an orientation on product quality and adding what consumers perceive as more value to product (Dubbink *et al.*, 1994). This innovative orientation is often termed a quality or value-added strategy.

The desirability or necessity of developing a quality strategy for the fish chain at large is commonly accepted. Sometimes individual firms innovate: fishermen take care to land fresh and undamaged fish, processors innovate and develop new products for the convenience market and high quality catering market. However, it seems difficult to implement quality and value-added improvements on an industry-wide scale. The individual, small-scale and short-term orientedness of the market structure in fisheries seems to prohibit the development of chain wide strategies for increase of the performance of the various parts of the chain as a whole. A gap is observed between the 'vision', making use of intrinsic qualities of the fish to position it at the top of the food market, and 'reality': a sector in crisis, uncertain about its future, conservative in its business strategies and unwilling or unable to change. For some observers of the sector it is disappointing and leaves them pessimistic about the prospects for a type of change in which there is room for fishermen as independent entrepreneurs. The dominating actors in the chain, the multinational food industry and the big retail chains, lead current innovations in the sector. To enforce innovations in the sector, they strictly control their suppliers on quality and punctuality of delivery. As a result, the trading/import/export companies (or the fillet enterprises) that supply the fish to the final processors demand high quality. Therefore the fishing industry is not only confronted by 'political' command-and-control quality, safety and health regulations, but also by the top-down, centralised chain management strategies enforced by powerful economic actors in the fish chain, located at the consumer end of the chain.

It is often, if sometimes reluctantly, suggested that because of the inertia of the fishing industry, the innovation strategies 'from above' are the only ones feasible and practically possible. Because of the short-term, individualistic nature of the market and the businesses and people involved, the fishing industry seems incapable and unwilling to make the necessary changes on a voluntary basis. Therefore it is suggested that there is a need for the intervention of dominant economic actors that organise the chain on their own terms and supervise the small businesses that supply them with the necessary primary or semi-finished products. The core company's strategy is to seek indirect control of the chain, not by vertically integrating all the chain activities in its own plants (the Fordist strategy), but in a structural way in which legally independent partners are steered by order portfolio, control and support on necessary technology and loans conditions. This form of organisation and management of the value chain is often referred to as 'Toyotist', after the control strategy of the Japanese car producer (Ruigrok and Van Tulder, 1995).

In opposition to the views that centralised forms of chain management are inevitable, when modernising the fish industry, we shall argue that sector-wide innovation in which fishermen and adjacent processing firms can survive as independent businessmen and not only as dependent suppliers, might still be possible. As an alternative for organising the fish chain along the principle of mass-production, we think it is possible (although far from easy) to organise parts of the chain on the principles of 'flexible specialisation' in which small and medium-sized independent firms compete and co-operate in networks (Piore and Sabel, 1984; Storper and Salais, 1997) or in other ways create a much more efficient information transfer between the links in the chain than is the case today.

This does not to deny the reality of internationalisation of the economy and, therefore, the salience of companies that represent this internationalisation. It is quite reasonable to assume that without a multinational food and retail sector, a quality and value-added consumer market would never have appeared. However, we think that it is beneficial for society if the power of the companies at the consumer side of the chain is balanced by an adequately organised network of Small and Medium Enterprises at the harvesting side of the chain. The future survival and prosperity of the industry depend on the ability of its players to exploit the new opportunities at the quality end of the market. This can only be achieved by shifting the orientation of the sector from production and volume towards demand and quality. A marginalised and alienated fishing sector is, in our view, not geared to such a re-orientation.

On the basis of recent theories on the role of information, knowledge, trust and co-operation in economic processes, we show that opportunities for quality/value-added oriented innovation can be created when various parties in the chain co-operate in networks. This co-operation is not something that happens automatically, it should be organised. A quality-induced innovation strategy within the fishing chain needs co-ordination between the various parts of the chain: in many, but not all aspects, the quality strategy is a collective effort of the chain and therefore needs to overcome the well-known problems of collective action (Olson, 1965). Co-ordination within the economy can be done on the basis of market signals and of centralised, hierarchical rules, but also by way of co-operation in (flexible) networks of independent firms. All too often it is taken for granted that the necessary co-ordination in the fish chain can only take place in the form of vertical integration under centralised governance of a few core companies. Therefore, in this contribution we concentrate on the opportunities for network-like, interactive forms of chain governance.

First, we discuss the changes taking place in the European fish chain and its consequences for different sectors in the chain. Then we ask ourselves who should take the lead towards quality improvements in the fish chain. We shall see that currently most initiatives are taken in retail and leading firms in the processing industry. Afterwards we zero in on the discussions for improvements in the fish product chain that could be taken in the harvest-end of the chain. We conclude on the importance of organising co-operation in the fish chain and develop some proposals to bring this about.

The Changing Fish Chain

The traditional neo-classical economic theory mainly deals with standard products with relatively coarse grades may define quality differentials. Open-market transactions, in which price changes moderate fluctuations in supply and demand, predominate.

The slowness of open-market mechanisms in communicating appropriate signals to producers and the lack of consistency in producer response has led to an accelerated movement toward greater integration of the agribusiness in the US (Johnson, Petrey and Schroter, 1996).

After a long period of time in which the research of innovation has been preoccupied with the technological information and technological change as the dominating theme, there are many researchers with an increasing interest in other forms of innovation like social, organisational and institutional innovation. A growing literature on these fields and on themes like the general function of knowledge and information in the economy suggests this interest (OECD, 1993). Some even speak of a new historical era where the economy is more strongly rooted in the production, distribution and use of knowledge than ever before (Foray and Lundvall, 1996).

Over the last decade there has been a focus on the importance of knowledge and learning for the development of new institutions as a consequence of the development within the new institutional economic theory (Lundvall, 1988, 1992; Cornish, 1995, 1997; Arrow, 1984; Storper and Salais, 1997). The research about the user-producer relations argues that information and knowledge about the market, customer needs and wishes are valuable and indispensable to producers when creating new products (Von Hippel, 1988, Lundvall, 1988). Lundvall terms the continuous interaction between producers and their customers that results in producers learning from users and vice versa in an unstructured and informal process 'learning-by-interacting'. Research on user-producer interaction makes it clear that inputs of information about the market leads to more successful products.

If the inputs like market information serves as an important contribution to the successful development of new products, the process by which such information is found is also important for economic growth and ought to be investigated in order to understand the dynamics of the production systems. The communication with the markets consisting of effective co-operation with the users is one of the critical factors setting successful products apart from failures.

The function of market information is to identify products or product quality as most demanded and valued by the consumers in order to concentrate the efforts to create new products exactly where the largest potential of the market exists. *Such continuous information back from the market is a key input allowing continuous innovation.*

The traditional approach to describe and to analyse the changes in the European food supply has been to look at the changes within the different industries in the primary, secondary and tertiary sectors. In order to see the changes from raw material to the end-user and the interconnections between the sectors several authors have suggested that the changes should be seen as a system or as a chain in order to understand the changes in the production in the context with innovation and power relations (Marsden, 1996; Marsden and Wrigley, 1996). Again other authors concentrating on the exchange of commodities and commerce within the food sector have chosen to use the concept of Market Channel approach (Johnson, Petrey and Schroter, 1996). An important background to look at the developments in a chain perspective is the growing interdependence between the sectors in the perspective of change.

Changes in the Market

Until the mid-1980s, the orientation of the fish chain has been decided by the consumers' desire for cheap and sufficient amount of food combined with the producers' interest in and possibilities for satisfying this need. Over the last decade, a growing share of the demand in the food market, like in many other market segments, has been still more differentiated. Still more of the food production requires 'tailored' raw material with certain qualities. One of the great challenges for the links in the chain is to create food chains where every single actor in the chain acts in the context of the chain and the market and understands that every link in the chain is of vital importance as a precondition for a production of high-value products to succeed.

When we speak of a strong trend of buying improved quality and greater product differentiation, it is important to mention that it does not mean that all standard products are disappearing from the European food market. Firms need to choose between a cost strategy that allows for price competition or a differentiation strategy in which quality (added value for the consumer) is the decisive competing factor and margins per unit of production are higher. According to Porter (1980) one should take care not to get 'stuck in the middle' of market polarisation. At the same time when consumers demanded raising qualities, retail chains offering discount products boomed. But the standard products are much more exposed to strong and increasing price competition, as import from third countries is likely for many standard products. For a producer of standard products there are only a few ways to survive. The most important way is to minimise the cost of production per unit, by creating economies of scale. Therefore a cost strategy is often connected to increase production.

Recent Changes within the Industry

In Europe the changes of the market situation have not led to increased vertical integration for the fishing industry in terms of formal ownership (with a few remarkable exceptions in Norway, Spain and Germany). The key developments in the fishing industry are a massive introduction and use of new technology and electronic equipment in navigation, catch searching, telecommunication and fishing gear. This development has created high demand for new knowledge on the part of fishermen. The majority of these learning processes has taken the form of learning-by-doing (Arrow, 1984). The most important trajectory for the fishermen has been to catch more fish using less manpower and more technology.

A general problem for the primary producers is that prices of products from farmers and fishermen are squeezed because:

➢ There is heavy price pressure from the other and more powerful links in the chain, especially the retail chains.
➢ There is very little direct contact and exchange of information between producer and retailer.
➢ Fishermen and farmers still produce bulk or standard products. They need to produce much more differentiated (tailored) products.

As a result, 'For at least a decade, the prices of agricultural products have tended to fall on international markets while those of manufactures have tended to rise.' (FAO, 1993: 51).

The share of the price paid by the end-user is generally reduced for the primary producers, as the end product becomes more manufactured. But even without changes in the level of manufacturing the primary producer's share is falling. This is a trend the producers have tried to prevent through rationalisations, economics of scale and new technology.

Fishermen have introduced new gear, electronic technology to cut down on search time at sea. Without cutting the fishing fleet the Danish fishing industry increases efficiency by 2-4% every year (Vedsmand, 1998).

Every single food chain consists of many independently managed units, all of which seek to maximise their earnings through the highest possible prices while minimising costs. Although, considering the market developments, the introduction of innovations that combine the reduction of costs with the increase in production is rational from the perspective of the individual fisherman, the collective result of these individual strategies is the well-known situation the European fishing fleets now find themselves in. The creation of a technically efficient fleet has resulted in a de facto capacity that exhausts the fish stocks as well as the profitability/viability of the fishing industry. The trend for fishermen to seek survival by catching more fish has been stopped, but it is still this trajectory the fishermen and their organisations are geared to.

Changes in the Processing Sector

The EU fish processing industry has a work force of approximately 78,000 workers. In olden days, the food processing industry delivered their products to the fragmented, self-owned retail stores. Their strong brands permitted them to set the conditions for everything from terms of payments to new products. Through their costly consumer investigations, they gained vital knowledge on consumers' buying habits and desires for new products, and had the capacity to innovate and develop new products and the responsibility of the presentation of the products in the groceries. However, the large retail chains successfully challenged the processing industry for their knowledge ownership in the 1970s.

Processing companies can roughly be divided into two main types:
➤ The very large factories, producing standard products such as breaded fillets or fish fingers. Most of them do not use the raw material caught by EU fishermen any more, but cheaper species of fish like Alaskan pollack and hake, imported from third countries. These factories face competition from new processing plants in Poland, Estonia and China and elsewhere.
➤ The more flexible processing plants, processing specialities, smoked products, and fresh skinless and boneless fillets. They are largely dependent on fresh local supplies. Most of them are small family-owned companies with few resources to change their production and to innovate.

Changes in the Retail Sector

For over thirty years the most dynamic link in the food chain has been the retail sector, taking over the initiative and power from the processing sector. They are 'the new masters of the food system' (Flynn *et al.*, 1994). With their concentration of buying power, the retail chains have forced the producers and the processors to give discounts, produce and distribute according to specifications given by the retail chains to keep their products on the shelves of the supermarkets.

The large retail chains are able to dictate their terms to the suppliers, from primary producers to processors. The retail chains are also in charge of the contact between the consumers and the food producers and processors. The large retail chains collect and utilise information on the behaviour of the consumers via electronic information in connection with their points of sale. This knowledge is utilised to take decisions on innovation and new products, often in their private-label product series.

In the beginning, in the early 1970s the private-label products were low-quality and products were introduced to compete with the leading brands on prices. At the start of the 1980s the leading retail chains radically changed the image of their own label products into high-quality and innovative products, successfully challenging the leading processors' brands.

This meant a major change from a situation where large brandname processors in the manufacturing industry developed their products to a situation with the retail chains utilising their knowledge to order new products accepted by the consumers in networks of sub-suppliers or even by the leading processors under the name of the retail chain. Thus the governance relations in the food chain have changed fundamentally. Especially within a range of products such as chilled ready meals, which are totally dominated by private-label products, innovative production has appeared as interactive relationships in long term co-operation. However, they are partnerships that are dominated by retail.

Private labels are developed in sub-contracting relations. Thus the composite character of the manufacturer-retailer interface is one of dynamic complexity. These networks are created by proactive retailers and are characterised by intensive interaction (Doel, 1996). An important reason for the continuous changes has been 'changing consumer tastes towards convenience, health, and indulgence' and the concentration tendencies within internationally operating food and retail corporations. The most expanding product groups are those which tend to require intensive interaction throughout the supply chain. Since the mid-1960s Britain's successful multiple food retailers have certainly evolved into 'innovative organizations' (Lazonick, 1991).

Who Can Take the Lead?

We are still in a situation in which none of the main actors alone or together have managed to create the necessary coherence to satisfy European consumer needs and demands for high-quality fresh fish products. It seems that a sequence of changes is needed, stretching through the whole chain from fishermen to consumers, shortening overall turnover time for fragile fish products. These demands radical changes in the traditional first-hand sale systems and logistics in order to save time and in a way giving possibilities to identify the fish through the different links. Also a further differentiation of the quality and prices is needed. Furthermore it is important to change the systems of first-hand sales in a way that the fish is no longer anonymous. New possibilities to identify the fish from catch to table and provide the products with reliable labels identifying the fish with time of catch, vessel and gear and labels indicating whether the 'cold line' has been broken in the chain.

These are some of the demands the consumers came up with in a recent "consensus conference" held in Copenhagen about the future in the supply of fish. A segment of the consumers were even more demanding on environment labelling. (Teknologi-Rådet, 1996; for further documentation of these trends, see also Grunert *et al.*, 1996). The way the system for fish products is organised today, it is impossible for a single actor in the chain alone to secure such changes in the supply of fish but the problems must be solved in some kind of collaboration between the main actors of the system.

But who should take the initiative to fulfil the tasks? A growing amount of literature on institutional economics is now dealing with the question of how the institutions governing society are changing or unchanging. North argues that institutions in society are not chosen for their efficiency but are the outcome of traditions, habits and practices and actual structures of power. 'Institutions are not necessarily or even usually created to be socially efficient; rather they, or at least the formal rules, are created to serve the interests of those with the bargaining strength to create new rules' (North, 1997).

North characterises institutional changes in five flashpoints:
> The continued interaction between institutions and organisations in the economic processes that make competition the key to institutional change.
> Competition forces organisations to continuously invest in new knowledge

and skills in order to survive.
> The institutional structure dictates which skills and knowledge are perceived as necessary to obtain maximum outcome.
> Cognitions are obtained from actor thought constructs.
> Institutional changes are continuous and path-dependent.

Thus the processes of change may stem from endogenous and isogenous processes, but the fundamental source to institutional changes is the learning processes of the actors of the organisations involved. Minor segments from each of the main actors in the food system for fish have already tried to solve the problems and some of them have succeeded within their niches to establish the exchange of information. But the chain is still waiting for a real breakthrough.

Three key links in the chain can be discerned as possible initiatiors.

Retail

Nowadays retail seems the most active in promoting change and innovation in the fish chain. It is the part of the chain with the largest experience to organise radical changes and to innovate organisations and products and has financial as well as knowledge resources to bring about changes in the chain. Besides the retail firms have plenty of information about customer habits and their actual choices, and are in the position to influence these choices by price setting, advertising and specials. The retail sector revolutionised the meat sector many years ago, so why not the fish sector? There are signs that the large chains are moving towards closer connections and direct contacts with the fishing industry.

Fish Processing

A large sector of the European fish processors are large-scale processors using frozen intermediates, semi-processed standard products from third countries for their production. The other part of the processing industry producing quality products has not changed much recently except for the required certificates, possibly caused by insufficient direct information exchanges with the consumers.

The outstanding exception is Unilever. The giant has recognised that something extraordinary should be done and that no other organisation was up to the task. Unilever took the international initiative to create a new institution called Marine Stewardship Council (MSC) in 1996, in collaboration with the World Wildlife Fund for Nature. Even from different interests they could agree on the same goal, to develop more sustainable fishing practices. The goal with MSC is to establish an independent institution with the task to carry through an operationalisation of a certification process enabling to label fish products caught in a responsible manner. It is the intention to include the handling of the fish in the chain to the consumer in this process, which should be completed in 2005. Three of the largest UK chains, Sainsbury, Tesco and Safeway have announced their interest in and support for this initiative. The participation in the MSC process is voluntary. But it is the hope and expectation of MSC that consumer demand for such environmentally certified products will grow to a point that the European fishermen and retailers have no choice but to catch, process and retail the eco-labelled products if they want to stay in business. However, the effectiveness of private initiative eco-labelling cannot be taken for granted as experiences with the certification process in the Forest Stewardship Council has learned.

The Fishing Industry

The European fishing industry has a basic advantage over fishermen from third countries in relation to the their home market without yet being able to fully utilise it. The advantages for the European fishing industry have a potential for better knowledge and information about the demands at the market in its own area of culture. Moreover it has the advantage of location, shorter time delays from the fishing grounds to the market including possibilities to serve the market with more fresh products than the third-country fishermen that already dominate the frozen-product market. There are examples of how segments of the European fishing industry have utilised their knowledge or intelligence in responding to market signals (see Chapter 6 of this volume and Friis and Vedsmand, 1998). But it looks as if the fishermen as a group are not yet adequately organised and lack the collective ability and willingness to adapt to the changed market circumstances.

Opportunities for an Interactive Fish-Chain Organisation

The brief survey above indicates that currently most initiatives for quality-induced innovation are taken in the links nearest the consumer end of the chain. The harvesting sector seems unprepared, and therefore unable to cope with the changes in the food market. The marginalisation of the fishing sector made it quite unfit to change its orientation on volume production to a more demand-oriented strategy. In our opinion this lack of preparedness is not a 'natural', unchangeable fact. Also, a fishing sector that is unable or unwilling to 'modernise' creates severe limits to the innovation strategies of other parts of the chain, as it is still the fishermen that have to take the fish out of the water. A confident fishing sector that is modernised in a technical as well as an organisational sense can become the sorely needed countervailing power for the core companies in retail and food processing.

Economic growth and increasing economic performance in the fish chain can no longer be based on increases in the volume 'produced' (i.e. caught) due to the inherent ecological limits of the sea. Therefore, economic development can only be based upon an increase in the added value to the product per unit of fish. The fish chain and all its elements should be organised in a way that per unit of caught fish the final consumer is willing to pay more. The quality strategy seems to be a win-win strategy in which a surplus is created in which every part of the chain can realise gains. That is why many wonder why fishermen do not seem very willing to go along with quality measures. Two reasons come to mind:

First, as a result of favourable circumstances at sea and the market in the first twenty-five years after the Second World War, an orientation on volume growth arose in all parts of the fish chain. When, due to alarming reports on the state of various fish stocks, harvest-limitation regulations started in de the seventies, fishermen became the prime and only subjects of regulation. In the dominant models on overfishing, the behaviour of fishermen was considered the main cause of stock depletion and, as a consequence, this was what scientists, policy makers and administrators focussed on. On the other hand, technical innovation continued and real capacity expanded (sometimes even stimulated by national or European innovation subsidies/structural funding). Under economic downturn, the socio-political environment was inclined to regard fishermen as culprits rather than as victims themselves. In return, fishermen saw themselves as victims of economical powerful actors (who did not offer them a good price) and a dysfunctional political system (national and certainly European!, whose harvest regulations made it impossible for them to make a honest living for their families), and 'forgot' their own contribution to stock depletion. In the end, many are more volume-oriented than ever before, as they considered that, under these highly uncertain circumstances, the biggest boat would be the best guarantee for continuing their business, and the best thing to do would seem to be catching as much fish as quickly as possible.

Second, a quality strategy in the fish chain is a long-term strategy: costs tend to be incurred first, while possible profits are reaped later. A quality strategy is also a collective strategy: it does not make much sense to invest in one part of the chain if other parts are not co-operating. Furthermore the link in the chain where investments are made is not necessarily the place where profits are reaped. Therefore, participants in a quality strategy need to have trust and certainty that their investments will pay off in the future. Under the circumstances as sketched above, fishermen are quite sceptical on any initiative that wants to intervene in their way of doing business, certainly if it is not (made) clear what's in it for the fishermen themselves in the short or long run. Certainly in chains, centrally managed by leading firms in retail or food industry, in which fishermen are no more than dependent suppliers, there is good reason to believe that the surplus created through a quality strategy will not be transferred to the fishermen.

Brief, both factors contribute to the fishermen's disbelief and distrust in innovative proposals that come from 'above'. Therefore the harvesting end of the fish chain should be stimulated to go along with a quality strategy. However, the harvesting side should not be forced by powerful political or economic actors to go along but fishermen have to organise themselves to adopt a quality strategy and at the same time to create a countervailing power to balance the multinational firms at the demand end. In this way they can contribute to innovative strategies of demand side actors for the benefit of all, and on the other can create a collective position in which fishermen can bargain for a fair share of the increased overall turnover.

The basis for the bargaining position of the harvesting sector is the control and quasi-ownership fishermen have over the fish stock. After some twenty-five years of regulation of harvesting levels in Europe, most if not all fisheries have become 'closed shops'. Due to a variety of quota systems, license schemes, capacity measures and other regulations, there are huge barriers to entry: it is quite difficult to become a professional fisherman these days. This closed-shop or cartel situation is certainly not without its problems. The role of capital invested in quota rights and licenses has grown. It has become increasingly difficult for younger men and women in fishing communities to become independent fishermen. The problem of succession of family members has increased (due to the capital involved, a son that takes over the business of his father starts with a huge debt to the other family members). However, the cartel situation results from the entry limitations that make sense from a stock conservation point of view and could form the economic basis for a strong bargaining position vice versa other parts of the fish chain.

Fishermen as a collective have not made effective and socially productive use of the fishermen's cartel till now. However, there are reasons to believe that in some European countries fisheries are now sailing more tranquil waters now that regulators as well as fishermen, after twenty-five years of regulation, have adapted to the fact that 'the sea is no longer free'. The Dutch flat-fish industry is an example. The Biesheuvel group regime has prepared the way for fishermen to make collective use of their position as quasi-oligopolists (together they control 75% of North Sea sole and 28% of North Sea plaice). But only if this cartel is properly organised can the whole chain profit. If fishermen use the cartel to maximise short-run profits, the chain as such will not benefit from it and in the long run, contradictions and tensions in the industry are bound to return as buyers look for substitute suppliers. However, fishermen could also use their combined strength, for example to limit catches of undersized fish, to realise market-information systems and fish planning systems and improve the quality of landed fish.

The theoretical basis for a strategic re-orientation of the fishing sector can be found in the concept of flexible specialisation, as developed by Piore and Sabel in their book *The Second Industrial Divide* (1984). 'Flexible specialisation is a strategy of permanent innovation: accommodation to ceaseless change, rather than an effort to control it. This strategy is based upon flexible, multi-use equipment; skilled workers; and the creation, through politics, of an industrial community that restricts the forms of competition to those favoring innovation' (Piore and Sabel, 1984: 17). The concept of flexible specialisation, taking the industrial districts in Northern Italy as examples, is a plea for industrial strategies based upon economies of scope and craftsmanship in a modernised form, instead of economies of scale and mass production.

The concept of flexible specialisation seems to fit especially the parts in the fish chain that consist of many Small and Medium Enterprises (SMEs) that vary in size but where no firms are able to assume a position to dominate others. As these firms are often locally concentrated, they are in the position to co-operate in many ways and at the same time remain flexible enough to react on market changes with alacrity. It is known however, that just because of the relatively small scale of the firms involved, flexible specialisation networks risk to become easily dominated by bigger firms further down the chain. The individual bargaining position of small firms vis-à-vis bigger firms that often are in an oligo- or monopsonist situation, is small. Therefore networks of SMEs have to be supported by various kinds of collective institutions. To develop a collective strategy the fishermen are in need of collective institutions in which co-operation can materialise. As is often the case in sectors dominated by small- and medium sized enterprises as in the agricultural sector, there is a rich variety of supporting institutions. Some of them are voluntary co-operatives and others are semi- or fully public institutions. Some of the institutions to defend the interests of fishermen are now under siege, because it is believed that they impede the market process. In our opinion, transforming these institutions is a more promising strategy than abolishing them. When these supporting institutions are transformed, they can support the modernisation process of the fisheries sector. These institutions will channel the current social and market changes such that the fisheries sector are not flooded by these changes but will be empowered to participate fruitfully in the current changes in the European food market.

We think that the re-orientation process within the fisheries sector is aided by the following three institutional innovations:

1. A Greater Economic Role for Fishermen's Organisations

In a more balanced chain, the market power of multinational food and retail industries, based on the knowledge of and control of consumer demand, is countervailed by the market power of fishermen, who have knowledge of the harvesting process and consider their rights to exploit the ocean's fish stocks as assets in a collective enterprise or shares in a joint-stock company (Van Vliet, 1998).

The options for a productive use of the collective control over the resources fishermen have not been put into practice until this day, since the co-operation between fishermen to realise these options, has yet to be organised. Under the uncertainty of the recent twenty years, most fishermen have opted for an 'individualistic' approach in their way of doing business. That is not to state that fishermen would not co-operate at all. Fishermen often co-operate in selling inputs cheaply, such as fuel and fishing gear. Fishermen also co-operate in defending their interests towards the political system and the government. In some cases the fishermen's representative organisations have lobbied quite effectively. However, when it comes to deciding what, when, where and how to fish and where and when to land and sell the fish, fishermen often steer their own course. However understandable, the individualistic decision-making of fishermen leaves them collectively vulnerable to changes in the market and to the strategies of post-harvest actors. Therefore we recommend that fishermen themselves try to organise their countervailing power through their established organisations. This can be done in several ways.

Fishermen's organisations could play a more extensive role in adjusting the decision-making processes concerning the operations on a daily, weekly and monthly basis. This is already done in some cases in the development of market information systems that possibly could evolve into a kind of fish planning system. With the help of modern electronic data transmission systems the knowledge fishermen have of market and price developments is made as up-to-date as possible. On the basis of the most recent market knowledge, fishermen could be advised (or pressed) to change their fishing plans. Because fishermen are still often hampered by a lack of knowledge on market developments, a good organisation of relevant market information by fishermen's organisations, is a relatively easy way to improve the performance of the sector. However, it must be recognised that improvements in market intelligence systems leading to adjustments in fishing and landing behaviour are an essentially reactive instrument, primarily meant to follow the market as quickly as possible, not to an instrument to change the market structure by collectively gaining a stronger position in it.

Fishermen's Organisations could play a more central role in influencing investment decisionmaking of fishermen on an individual as well as a collective basis. For sound reasons, individual fishermen are inclined to plough back earned profits into the sector itself. By increasing the capacity of the vessel, or buying an additional one for an heir, fishermen take care that they themselves and their offspring have a future in fishing. Although this makes sense from an individual perspective, at the collective level it leads to the unwanted result of too many vessels chasing too little fish, and therefore to an increase of the fishermen's race for the fish. Now it is increasingly difficult to expand in terms of harvesting power, there seems to be a task for the fishermen's organisations to direct new investments into the post-harvesting sector like on-board packaging of fish. The organisations must make clear that investing in post-harvesting activities is not a withdrawal from fishing but an offensive action, directed at job creations and sustainable wealth for the fishing community because investing in the post-harvest channel gives fishers influence on a greater part of the value chain.

2. Re-orienting the Auction System

The public auction system, once developed to protect individual fishermen against traders with far more knowledge on the market and prices, is now often considered an obstacle to information transfer, to the development of stable price levels and to direct, long-term partnerships between individual fishermen and individual processing firms. Therefore some foresee the closure of auctions in the near future. However, despite signs of concentration in harvesting, in important parts of the market, such as demersal fisheries the typical fishing firm will be relatively small in comparison to traders/processors. There will be information gaps between fishermen and traders. As a result, there remains a need for a mechanism in which information of quantity of fish landed and prices that are paid remain publicly known. A public auction system remains an important institution in the fish market.

But auctions need not to stick to a single trading mechanism: the clock auction or other types of floor auctions. A new type of auction system is developed and now tested in IJmuiden, the Netherlands. The main objective of the Infomar project, run by a privately-owned consortium and partly funded by the EU is to create a real-time information highway linking fish buyers, sellers and fishing vessels. As a result, all members of the fishing community will benefit not only from data as to what is available on the market but - because of the link with the vessels - what will be available several days onward. With this advance indication of both supply and demand, buyers can plan their purchases and sellers can arrange to get their product to the market where its profitability will be highest. The result should result in higher value for the fish on the market at a more stable price, less waste and a more efficient use of a natural resource.

The main task of the Infomar project is to define, develop and run as a demonstration project, an electronic inter-European information and trading system for fish products, especially suited for fresh fish and fish products with short life cycles. This will enable the seafood industry to reap the benefits from this new technology in a way that other branches of the food industry, such as production and retail, already do. Many obstacles have been identified hindering this development, but solutions to all of them now exist. The task is to adapt these solutions and integrate into a single system.

As far as the system goes, an Infomar type system has much to offer. There will be no single system of fish sales and marketing in operation at any time in the future. What will exist is a range of options as to how producers go about marketing their catches, how merchants, processors and wholesale and retail organisations source products and influence, and how they influence the mode and timing of harvesting, handling and distribution. Infomar offers just another dimension to this equation, complementing existing trading systems - shout and electronic port auctions, the various forms of supply agreement, the operations of vertically integrated structures - and information sources – the trade press, specialist market information providers, auction houses, POs etc.

3. Financial participation in innovative market developments

Too often fishermen see no other option but to plough back profits earned into their own vessels, in catching capacity and quota rights. It is questionable that these investments are the most profitable with a view to the future of the fishery as a whole. It consolidates the fishermen's dependence for their income from the volume caught. In the long run it seems more profitable when profits are used for investments in the post-harvesting part of the chain. Future value creation will not arise from increased fishing capacity, as the sea's resources are limited. Instead, increased investments in harvest capacity can be expected to lead to increased competition between fishermen. Investments in adjacent industries, however, will strengthen the interdependence in the fisheries 'business districts' around the ports and auctions. Furthermore it will create employment opportunities for sons and daughters of fishermen are no longer able to go out to sea owing to catch limitations.

Although fishermen understand the logic underlying the argument, they are often reluctant to invest in processing. On the basis of past experiences, some argue that it is impossible to manage a vessel and a processing plant at the same time. And of course you need to be in control of a firm in which you have invested funds. Others are afraid to lose control of the money invested. These feelings are based upon past experience and quite understandable: fishing is a different business than trading, processing or exporting fish. If you don't know these markets, it is risky to invest your money in it. However, it seems that with help of co-operative banks etc. these hesitations should be overcome, certainly when younger generations enjoy better formal education.

Conclusion

Profound changes need to be realised in the fisheries sector, with a view to focussing on consumer demand for high-quality, secure and sustainably harvested fish and aligning the sector's and consumer understanding and definition of quality. We argue that these demands will be met with or without changes initiated by the fishing industry. The question is which link (or links) in the chain will take the initiatives to meet the demands from the consumers.

We found that the retail link has been the most proactive innovative force and has gained the greatest experience in bringing about major changes. It is our hope that the fishing industry itself will act in a much more proactive way in the near future for at least two reasons. First, because it is the experience from other sectors that all links in the chain must acknowledge the changes and the goals of the changes, not least primary producers. Second, we argue that the fishing industry must take part and actively co-operate for such changes if the industry is to survive as an independent industry and not as dependent suppliers to other parts of the chain.

11 Balancing Boundaries

Towards a Regionalised Fisheries Governance System in Spain

JUAN-LUIS SUAREZ DE VIVERO

Introduction

This Chapter will discuss opportunities for a more decentralised system of fisheries management in Spain. The Spanish governmental structure, based on a division into seventeen self-governing Autonomous Regions, provides good conditions for handling the great diversity in fisheries organisations among and within coastal districts.

The existence of a great variety of fisheries organisations is also conducive to involving user-groups in the decision-making process. Nevertheless, co-ordination at the national level is a complicating factor, and so is the co-ordination of fisheries policies between the regional and the EU level. This may very well create a situation where regional autonomy and fisheries co-management is more symbolic than real.

Organisational Diversity

The different organisations operating within the Spanish fishing sector can be grouped under five basic types: Fishermen's Guilds (FGs), Shipowners' Associations (SAs), Fishing Producers' Organizations (FPOs), Business Associations and Trade Unions. Considering their functions, the SAs, FPOs and Business Associations can be grouped as representing the corporate side of the extracting, manufacturing and commercial sectors. The Fishermen's Guilds and the Trade Unions would form another sub-group with a mission to defend labour interests.

It is precisely in the functional aspects of the organisation where we find the main problems of the fisheries sector. The panorama consists of organisations with a very old structure, like the FGs, and newly created ones, such as the FPOs, derived from the Common Organisation of Markets (COM) established in Spain since it joined the EU. Of these five types of organisation, the FGs stand out owing to their ancient origin (some go back to the 12th century), composition (both shipowners and crew) and their importance within the sector. The FGs configure quite a broad associative network, that is, strongly established in the local communities and deeply rooted in the sector. Such characteristics constitute both a solid asset as well as an extra difficulty when faced with the new challenges confronting the fishing industry.

The associative scene shaping the Spanish fishing sector forms a structure of overlapping of functions and responsibilities between the five types presented here. Despite the diversity in organisational structure, their goals and tasks, most of these organisations have overlapping domains. This is especially the case with regard to the space occupied by the FGs, which is partially taken over by other organisations. On the other hand, if we examine the most important functions each has developed, we can see frequent intersections, the same tasks being undertaken by several organisations. It is only a logical result of the above that this is more frequent in the case of the FGs.

Government Administrative Organisation

Since Spain joined the EU, the administrative structure of the fishing sector is organised at four different levels: European, national, regional and local. These four levels have a pyramid structure as the flow of decisions goes from the top to the base, never the other way around.

The key level is represented by the European Fishing Administration, which sets different decision levels determining the bases for the functioning of the sector. The decisions taken at the EU and Member State levels set the guidelines of the Common Fisheries Policy and of State fishing policy respectively, where the Spanish state structures the regional and local level. The latter two levels are directly related to the political and territorial divisions of the State.

Within this administrative structure, the articulation between fisheries organisations and administrations is still weak, revealing a serious lack of participation on the part of the former. This underlines the non-existence of a body that would institutionalise and formalise the participation of the organisations in the decision-making process. As a consequence informal procedures are spreading based on the goodwill or understanding that exist in each situation.

For the organisations, the nearest levels within the administrative structure are the local and regional levels. This is due to the distribution of the associations within the sector, through their respective Regional and

Provincial Federations, and also to the dependent position some of them hold vis-à-vis the government administration. However, at the State level the presence and participation of the organisations is low for the aforementioned reasons. The FGs and the SAs have national federations that transmit their opinions to the central administration. At the superior (European) level, however, their presence and influence is nominal on the decisions that affect the sector. Thus, at this level the fishing sector Organisations have few opportunities either to face new challenges or to take on a more significant role on behalf of the fishing industry. Regional Administrations also need to strengthen their presence and influence at the European level, which would add importance to the region within the European institutions (Committee of the Regions). Another line of action already initiated by some of the Autonomous Regions is opening offices in Brussels, to make their interests heard within the EU bureaucracy, and to be part of the lobbying process at that level.

Regional Fisheries Policy

The degree of administrative decentralisation in Spain is strikingly high and results from the State's overall political structure, which is based on Autonomous Regions with political self-government - parliaments and autonomous cabinets. The management of fisheries falls to the nation's government as far as stocks and resources are concerned (except for coastal waters), whereas the regulation of the fishing sector is the Regional Administrations' concern. The political and administrative framework has definitely favoured the development of regional fisheries regulation. In fact, this was the model well before Spain became a member of the European Union. This model has not been without its problems, as marine areas delimited for management or research purposes such as the ICES divisions and other marine spaces known as 'maritime regions': the North Sea, the Irish Sea, the Bay of Biscay, do not correspond with any political or administrative regions that could perform a regional management function.

Therefore, the current political and administrative system only permits the regulation of the fishing sector in matters related to structure and markets, but not to resources, except for coastal waters. There are ten regional fishing administrations, one for each coastal Autonomous Region. This entails that a regional policy with a higher degree of development would count on an ample administrative base and experience in the management of the sector. The transition to a far more developed regional model would first imply increased authority over the resources of their neighbouring waters, that is, over the Exclusive Economic Zone corresponding to that particular region. This would create enormous problems as high-ranking precepts and regulations (Constitutions, Organic Laws) would have to be altered.

In view of the features of the Spanish fishing sector, with a large fleet and a large number of fishermen who have customarily been operating on international fishing grounds, one of the most important functions to be performed by the administration is the negotiation of agreements with other countries. This function relies entirely on the central administration, as foreign affairs falls exclusively on the State's power. In the current model of the State of Autonomous Regions, this seems to be one of the main limitations for the regional fishing authorities, especially in the case of those regions whose fleets work in other countries' fishing grounds. These regions are mainly Galicia, Andalucia and Pais Vasco (Basque country). The authority to sign agreements falls on both the nation's government and the European Commission, although the negative effects and consequences derived from the dependence on other countries, such as biological stoppages, arrests of vessels and the reduction of the fishing activity, are basically suffered by the regional administrations and the local communities.

Another likely mode of regionalisation, which has also been advanced elsewhere (Symes, 1997), is what could be termed centralised regionalisation. It consists of the delegation of the management of specific maritime fishing areas to a state or a group of adjacent states excluding inland countries. In view of the EU framework, however, this would clash with one of the most basic principles of the Common Fisheries Policy: the responsibility for fisheries management is vested in the nation state. In this regard, the management of North Sea resources, for instance, would be assigned to the United Kingdom, the Netherlands, Denmark and Norway as a group. Then, the process of regionalisation and the subsequent decentralisation would take place on a national scale, and the contents of the regional management would mostly affect the maritime areas delimited according to their vicinity to the coastal countries involved.

Model

According to this interpretation, a model of regionalised fishing management implies:

> a sub-national regional political and administrative structure with political authority on a regional scale and empowered to regulate the fishing activity at a specific level of competencies;

> (as a logical inference from the former prerequisite:) a national administration that devolves jurisdictions and co-ordinates the various regions to ensure a national dimension to the whole fishing industry;

> fishing areas linked to the political-administrative region (adjacent jurisdictional waters);

> decentralisation, understood not only as the transfer of administrative powers from the national to the regional administration, but as the devolution of functions from the Administration to the different groups of interest involved in the fishing industry.

This model has been partially introduced in Spain with the following strengths and weaknesses:

Strengths:

> The fact that the state's organisation is based on regional political autonomy.

> There are regional fisheries administrations with exclusive authority to manage resources in coastal waters (between straight-base lines and the coastline), the total surface of 14,400 km^2. Regional and central administrations share the ability to regulate the sector (structures and markets).

Weaknesses:

> A high degree of dependence on other countries' resources.

> As a result, the regional administrations lack the ability and the authority to face up to one of the most important limitations of the fishing industry.

> Scarce development of association networks. Despite the large number of organisations, including some traditional and long-established ones, such as the fishermen's associations, the need for cohesion, resources and managerial training is such that their possibilities of participation are underdeveloped.

Design Principles of Fisheries Management

In order to achieve a management system which would be effective for the whole Spanish fishing sector, I suggest a model based on three fundamental principles: greater regionalisation, decentralisation, and the involvement of all agents in fisheries management - that is: co-management.

Management Regionalisation: the transfer of specific powers and responsibilities from the central to regional authorities. Many competencies have already been delegated from state administration to the regional administrative entities. However the basic principles of fisheries management continue to be designed by the central administration, which, equally, is responsible for maintaining the unity of fishing policy which applies to the whole territory of the State. In order to achieve a greater degree of regionalisation it is necessary to expand the competencies of the Autonomous Regions, which are currently limited to the inland waters, to the other maritime areas.

Delegation of Management: the transfer of responsibilities pertaining to the formulation and implementation of fisheries management policies from government administration to the different organisations present in the sector. In order to establish a genuine delegation of functions, it is necessary to create a legal framework allowing the fishing organisations to assume new functions. It would also be important to create communication channels between these organisations and the administration. Just like decentralisation, the possibility of delegating functions is more feasible within an institutional framework that is not centralised, as is the case in Spain where the state is divided politically and administratively into Autonomous Regions.

Management Co-participation: the co-operation between the different groups and stakeholders taking part in the management of the fishing sector. Such involvement is needed since neither the government nor sector participants have the capacity to assume the management responsibility alone. First of all, it is increasingly important to address and moderate the conflicts of interests that prevail between the different administrative levels and the local and sectorial groups. Secondly, it is necessary to provide the organisations with the opportunity of real participation in the design of the most appropriate fishing policy that will affect them anyway.

Opportunities for Regionalised Management

In the particular case of Spain, the division of its territory into Self-Governing Regions, and the consequent administrative decentralisation, set up the foundations for a new model of fishing organisation, based on the regionalisation of management, allowing for the enlargement, deepening and development of Regional Governments' competencies. Such a proposal has the benefit of a widespread acceptance by Self-Governing Regions. The proposal provides them with an opportunity to develop their own management abilities and to realise for themselves the economic potential that this option harbours.

The rise of the region as the spatial base of fisheries management relies on certain antecedents of the Common Fisheries Policy, in which some guidelines referring to issues devolved to regions are interpreted and executed by regional bodies. Here, the State is left out of the process, although it is still responsible for the carrying out of competencies about international agreements. An example is POSEICAN, an EU-devised regional fishing board for the Canary Islands, to be implemented by the Self-Governing Region, but negotiated by the central administration.

Many regulations of the Accession Treaty of Spain to the EEC refer to aspects in which the Self-Governing Regions have exclusive competencies. On the one hand, current debate concerns the construction of the Europe of Regions, where the proposed model would have full validity, and where the principle of subsidiarity would serve as a starting point. Here, the EU would only assume the tasks that it might perform more efficiently than the member States individually. On the other hand, it is not easy to see how the role of the State as co-ordinator of these regions vis-à-vis the EU can be strengthened.

In addition, a model of fishing organisation where the execution of management tasks is vested in the different regional administrations, facilitates that specific needs of each region can be met more effectively. This is particularly important for a fishing sector such as the Spanish one, which is characterised by a great diversity of situations from one region to another with respect to type of fleet, level of development, share of population involved in fisheries, catch composition and the like. In turn, this enables a more effective organisation of the sector as it prevents the generalisation to all the country of decisions that concern specific situations.

This management proposal becomes more important when we bear in mind that the crisis of the sector has opened the door to new socio-economic phenomena, not currently considered in the decision-making processes, that condition the development of fisheries in the Spanish fishing communities, which are usually involved in traditional fishing modes. These phenomena are best addressed by a regional administration, which has a solid knowledge of the framework of fisheries, as those traditional sectors are linked to the informal ('black-market') economy, whose procedures and merchandising channels - characterised by their limited transparency - are difficult to quantify and analyse.

Problems of Regionalisation

In order to implement a regionalised system of management, a whole series of administrative problems caused by the conflicts and distrust concerning administrative competence issues has to be solved first. The State Administration considers the transfer of competencies to regional administrations a loss of power as well as efficiency. Despite its initial aims, the division of Spain into Self-Governing Regions has not always entailed more dynamic and efficient management.

As a matter of fact, a regionalised fisheries administration lacking channels of communication to the national level as well as the instruments to co-ordinate regulations for the national territory as a whole, might widen the gulf between the different administrative levels and the nation's fishing industry. This would make the present situation even worse.

Moreover, the problem of overlapping Common Fisheries Policies, EU policy on regional development and national realities is largely caused by the administrative morass of the Spanish regional autonomous organisation. In fact, the distribution of fishing competencies between the State and the Self-Governing Regions took place without an adequate planning of the fishing sector. Thus a match between the new political and institutional system was not fully accomplished.

Apart from the problems already described, which will emerge when the time comes to set a regionalised organisational system of fisheries in motion, and which are associated to such important matters as the need of changing the current legislation, we should keep in mind that management does not only concern issues related to regional fisheries. Often international agreements signed by the EU can be decisive for fisheries development at both regional, national and world levels.

For the present, the EU is responsible for the negotiation of international fisheries agreements; prior to that, the different member countries participate in discussions on a national scale. In this regard, it is difficult to plan a regionalised management system. Whenever a regional administration could negotiate agreements on matters such as fish exports and imports, it would have to deal with another country or with supra-national organisations instead of another region. This puts it in a clearly disadvantageous situation.

In conclusion, then, a management system based on the three organising principles described above carries obvious advantages. In many ways, these principles are also concomitant with the new EU philosophy concerning subsidiarity and regionalisation. But as any new administrative reform, it must face challenges in its environment. Decentralisation, regionalisation and co-management, separately or combined, will not take effect in an institutional vacuum. To work effectively, adjustments must be made with respect to the relationships and interaction between regional, national and EU levels. For instance, channels of communication must allow knowledge to flow freely between levels. Furthermore, the scepticism pertaining to such a reform for instance among central and regional government bureaucrats must be overcome. Their fear of a loss of cohesion of national fisheries policies is a concern that must be met. A major concern is also the fact that the extraction sector covers a wide range of fisheries organisations, fleets and fishing methods - from traditional fishing to the very sophisticated industrial freezing fisheries. The challenge here is to find ways to combine different interests and provide solutions to very diverse problems and situations.

Such reform also requires a clear delineation of competencies and responsibilities. Today, there is an horizontal overlap of functional responsibilities between fishing organisations at local and regional levels. A decentralised and regionalised management system risks creating overlapping responsibilities along vertical lines of authority. This can only be overcome if communication between administrative levels is well developed, and tasks are properly divided. There is a need for new patterns of interaction at local and regional levels as well as patterned interaction between levels of governance, from the local to the international. In every instance, the principle of co-operation is crucial for the effective performance of governance.

Based on the above discussion of the current state of affairs of Spanish fisheries management, the following suggestions should be considered:

➤ Expansion of the capabilities of the regional administrations in matters that are beyond their jurisdiction, such as the management of the resources and stocks of the territorial sea and the Exclusive Economic Zone.

➢ In those regions where the fishing sector is more developed, new administrative bodies (Regional Fishing Councils) with the intensive participation of the different groups involved in the sector can be introduced.

➢ Improvement of the communication channels between the central and the regional administrations. The division of jurisdictions has caused the isolation of the various departments and, despite the fact that there are shared jurisdictions, a spirit of inter-administrative co-operation is sorely lacking.

➢ The rigidity of the communication between the central and regional governments stands in the way of the relationships between the latter and Brussels. The central government acts as a filter between the European and regional level on the pretext that foreign affairs must be the central government's exclusive concern. As far as the regional governments are concerned, this blockage in the access of the regions to the European institutions is usually solved by creating 'delegations' in Brussels. Similarly, the fishing industry's establishment of lobbies seems to be an attempt to overcome the administrative filters on a regional and a national scale. In this regard, the progress made in European regional policy would have a very positive effect, as well as the encouragement of political bodies such as the Advisory Fishing Committee, which currently play a subordinate role.

PART IV
FISHING FOR
OPPORTUNITIES

12 Knowledge-Based Fisheries
Opportunities For Learning

SVEIN JENTOFT, PETER FRIIS, JAN KOOIMAN AND JAN WILLEM VAN DER SCHANS

Introduction

Fisheries policies are formed at the crossroads. Typically, they must address many concerns that are often hard to reconcile, such as resource conservation, economic development, regional differences, international relations, and the like. As concerns are many and complex, so are stakeholders and their demands. Stakeholders play no passive role in fisheries policy formation. In most European countries initiatives in the fishing industry are subject to hearings, negotiations, advisory procedures, and lobbying, thus involving a wide range of affected user groups. This makes processes of deliberation, implementation and enforcement cumbersome. Thus, the basic problem of fisheries is one of governance. It depends on the abilities of the many actors involved identifying common problems, of reaching a consensus on priorities, and, in the next instance, sticking to regulatory measures. Part of the problem is the lack of capability to learn - or agree on the lessons - from policies that misfired. Therefore, European fisheries are in deep crisis, as they have been for a long time, and the prospects for improvement look none too bright.

The exhaustion of the resource base and the excess capacity of harvesting technology, in part cause and consequence of over-harvesting, are only two of several urgent problems in fisheries. Fish is, often for good reason, associated with poor quality, inferior to other food products such as meat and poultry. Although fish has everything it takes to become a high-quality, high-price consumer product, the fishing industry is not always up to the task. Again we believe that this is a problem of governance and learning. Fish quality typically involves the whole production and distribution chain from harvesting to consumption. Quality deterioration starts as soon as the fish has left the water, sometimes even before that, when the fish is allowed to die in the nets. Later in the distribution process, quality cannot be improved; under the best of circumstances it may only be maintained. Since fish is often transported over large distances and

239

commonly bought and sold several times before it reaches the consumer, marketing is always a struggle with time. Therefore, product quality conservation is technically and organisationally challenging, and involves every actor at all stages of the distribution process. In promoting good quality, externally imposed restrictions and standards (such as ISO) can only do part of the job. Inter-organisational co-ordination and co-operation between trading partners, who must share and exchange information and knowledge, is key. At each step in the chain actors must know how their particular performance may affect final product quality. Therefore, not only rules and regulations regarding the technical aspects of quality matter, but also structures of inter-organisational relations.

The quality aspect may easily be subject to a Prisoner's Dilemma, where players take unequal shares in the costs and benefits of co-operation. As a result, no one regards it as their particular responsibility to contribute to quality improvement, as they have no guarantee that they will stand to gain from it. Similar to resource management, co-operation between chain participants may promote the realisation of high quality and better prices in the consumer market.

The good news is that efforts in enhancing the quality of fish products may also promote stock conservation. A stronger focus on quality would infer a shift away from the orientation on quantity as the fishing industry's prime income generator. The fishing industry and government agencies largely overlook the opportunity to 'kill two birds with one stone'. Quality management and resource management are typically regarded and treated as separate issues and may therefore easily work in opposite directions.

In this Chapter we shall discuss the relationship between governance and organisational learning at a conceptual and theoretical level, with fisheries as the empirical reference. We introduce a broad institutional perspective on learning as an integral part of the governance process and argue that institutionalisation of learning is crucial to governance. Also, for reasons mentioned above, we contend that this is a socio-political process involving a wide range of stakeholders and decision-makers. This is not only a question of substance, of what should be learned, but also of process: how to learn to interact effectively in the learning process. Learning must become a collective, interactive, inter-organisational process that concerns and qualifies the whole chain.

A Governance Perspective on Learning

The interatctive perspective on governance and organisational learning advanced in this book suggests that, since 'no single actor has sufficient action potential to dominate unilaterally in a particular governing model' (Kooiman, 1993: 4) governance is increasingly subject to forces of disintegration and strife, incapable of accomplishing anything more than fragile compromise. Under these circumstances, the scope for problem-solving and opportunity-creation at a collective level is slim. To become effective in problem-solving and opportunity-creation, governance must therefore ease collaboration and commitment among involved parties. Thus, governance is not only about problem-solving per se. It also involves the construction of the very social and cultural conditions on which problem-solving and opportunity-creation depends. What is necessary, then, is the provision of institutions, organisational structures and relations, and, no less important, an atmosphere of trust that induces partners to co-operate. Industries such as fisheries will not realise new opportunities for problem-solving where these conditions are non-existent or flawed.

This perspective of the shortcomings and potential of governance raises a series of interesting and important social research questions with respect to fisheries.

First, we know that some processes of governance are formally structured, highly institutionalised, and occur in settings open to public scrutiny. We need to understand fully how these structures restrict and enable problem-solving and opportunity-creation. Also, we need to realise how they enable or obstruct involvement of affected interests co-operatively, and how they shape the form and content of the management discourse. It is equally important to understand and describe those mechanisms and processes that are informal and take place backstage. Informal mechanisms and institutions may have a decisive impact on the outcomes, implementation and enforcement of governance.

Secondly, the policy process is partly one of social integration, partly one of aggregation (March and Olsen, 1989). In the first instance, preferences and viewpoints are deliberated and negotiated (as when user organisations and government agencies meet to discuss sectoral policies). In the second instance, they are registered and 'added together' (as when people vote in parliamentary elections). The former process occurs primarily within what Rokkan (1966) terms the 'corporate channel', while the latter takes place within the 'numerical channel'. The interactive model of governance emphasises that although institutional barriers exist between the two channels, of particular interest are the processes that occur at their interface. A shift in relative influence and power from the numerical to the corporate channel is problematic from a democratic point of view, and it is unclear whether this shift makes industries such as fisheries more or less capable of problem-solving and opportunity-creation.

Thirdly, the interactive perspective on governance challenges traditional perceptions of policy-making as a streamlined rational process through which goals are clearly defined and the most cost-effective means chosen. In contrast to the hierarchical model, the interactive model depicts social relations that are less structured. Participants form looser coalitions representing a diversity of interests, values, and worldviews. Here, they negotiate goals at every level and stage of the governance process. Goals are often vaguely defined, conflicting and difficult to reconcile.

These limitations on governance have several consequences: They make strategies inconsistent and outcomes difficult to predict. Organisations tend to have short memories, unstable preferences, and shifting ambitions. Agreements have limited endurance and they need to be continuously renegotiated. Deliberations and negotiations are inconclusive. 'Solutions' do not solve problems but change them - sometimes to the worse. Thus, by implication, the interactive perspective would not focus on policy outcomes as much as policy processes and the institutional framework within which they work.

Fourthly, problem solving requires a clear definition of the essence of the problem. However, the way actors structure their discourse may very well inhibit them. Even if participants can agree on the problem definition and compromise on what to pursue, they may still disagree on means. Means are not only technical; they reflect particular interests and perspectives. Frequently means are chosen because of their symbolic merits, for instance because they are easy to legitimise. They may have no real substance and be nothing more than 'window dressing'. Therefore, the discourse of governance, the transactions involved and their enlightening effect on the actors involved, are research issues of considerable interest.

Fifthly, the interactive perspective on governance holds out two likely outcomes: one characterised by social disorder, adversity, and possibly chaos; the other by innovation, creative energy, and constructive change. It would be interesting to research the circumstances under which one or the other occurs (or both). When does governance gain rather than suffer from the plurality of participants with different constituencies, experiences, perspectives, and knowledge? As Kooiman (1993: 4) points out, the problem is that 'no single actor, public or private, has all knowledge and information required to solve complex, dynamic and diversified problems, no actor has sufficient overview to make the application of needed instruments effective...' As a collective, however, actors may well possess all the relevant information and knowledge to fulfil the task. Yet to accomplish it, these resources must be united in collective action.

Pooling these resources may prove an intractable problem in itself. It would obviously help if actors are constituted as a group, and if incentives exist to encourage involved and affected actors to share their knowledge. Thus, interactive governance must allow for pooling of specialised competencies, but also for mutual, interactive learning throughout the decision-making process. Learning would then be both unilateral and multilateral. It would occur at all levels of governance, from the level of practical problem solving, which Kooiman in Chapter 1 of this volume terms 'first-order governing', to the second and third-order ('meta') level of governance. Bateson (1972) terms learning at the meta level 'deutero-learning' -, i.e. learning about learning. For the first level, Argyris (1992) speaks of 'single-loop' learning, 'double-loop' learning for the second order of governance. Single-loop learning takes place when mismatch between intentions and results are discerned and corrected. Double-loop learning happens when the basic variables and conditions that create disparity at the first level are identified and changed. Argyris argues that organisations should place greater emphasis on double than on single-loop learning:

> Although single-loop actions are the most numerous, they are not necessarily the most powerful. Double-loop actions - the master programs - control the long-range effectiveness, and hence, the ultimate destiny of the system (Argyris, 1992: 10).

However, that is easier said than done. Learning at the second-order level requires us to question and scrutinise fundamental assumptions and values. Such an exercise may be felt as threatening for organisation members who may be inclined to evade or resist it - what Argyris calls 'distancing'. Consequently institutional changes often occur only at the margins, new solutions are only sought close to old solutions, and they do not attempt to radically improve current affairs. Only major crisis is likely to cause more fundamental change. But even then inert forces are at play, such as powerful organisation members with an interest in maintaining status quo. Obviously, this is part of the diagnosis of the fisheries problem.

Interactive learning is a process in which participants learn from each other, and from each other's learning. Under the best of circumstances, these learning processes may occur at all three levels. For learning to become a permanent feature of the governance process rather than sporadic and ad hoc, second and third-order/meta learning is essential. It has to be encouraged and explicitly planned. This requires addressing some empirical questions: How are new opportunities for learning and problem solving discovered at each level of governance? Are new opportunities intentionally sought or are they primarily stumbled on? How is new knowledge fed back into the governance process? Does meta-learning occur, and how does it change governance and its institutions? The fishing industry is well suited for research along these lines.

Interactive learning requires systematic recording and reflection on experiences made throughout an organisation's developmental history. How organisations structure these exercises will determine the capacity of organisation members to learn and to share what they learn. It is not only a question of how individuals learn. The more challenging issue is how learning at the individual level penetrates the organisation so that it is preserved over time despite personnel turnover. Several inquiries are pertinent:

➢ How do organisations assemble knowledge, which and whose knowledge do they seek, and for what purpose is it used?

➢ How do decision-makers handle knowledge, how is knowledge accumulated and stored within the organisation, and how does it inform decisions at the systemic level?

➢ How do organisations involve participants in exchange of knowledge for interactive learning?

➢ To what extent is interactive learning routinised and institutionalised so that it becomes a continuing social process?

We can raise similar questions for entire industries: How do networks of organisations learn and adapt? Which characteristics of inter-organisational relationships are conducive of learning? To what extent do competitors and chain partners learn from each other, and how? How can firms co-operate in promoting a higher level of competence?

The problem of qualifying whole industries is structurally very similar to the common-pool situation in natural resource harvesting, such as fish (Olson, 1965; Ostrom, 1990). Sharing the natural common resource and sharing knowledge may both be in the collective but not necessarily the individual interest. Knowledge enhances one's competitive position, and although the knowledge itself does not suffer but may gain from being shared, those individuals that hold it may do so. If shared, knowledge may lose some of its value as a strategic resource for the stakeholder. How to overcome this problem by turning learning into a plus-sum game, as when people learn from each other, is therefore a research issue of great interest.

Usually, we must seek the solution in the incentives, institutions, and organisational structures that restrict, motivate and guide individual action. It often requires an external authority to act as a catalyst. In the absence of a hierarchical governance structure at the level of the industry, the government must often take on this responsibility. Also, only the government can establish some of the incentives and institutions necessary to promote collective action (such as tax breaks for educational programs). Yet industries and their associations can do much to promote closer ties between individual firms, thus easing interactive learning to the benefit of all parties involved.

Learning Opportunities in Fisheries

We can fruitfully address these issues of organisational learning with respect to fisheries. Having to rely on a fugitive, common-pool resource, subject to natural forces beyond human control, sets limitations of governance. Fisheries decision-makers live with high degrees of uncertainty due to unreliable data and crude analytical models of a seemingly chaotic environment. This makes future resource flows hard to predict and difficult to plan for. Given this, it is unsurprising that fisheries policies time and again have proven ineffective in addressing their basic challenges, such as protecting the resource base from over exploitation and securing viable coastal communities. One may safely conclude that if European fisheries policies were judged by the criterion of sustainable development there would be few, if any, success stories to report.

A complicating factor here is the extreme diversity and complexity of fisheries. The industry consists of units of variable scales and organisational forms. It ranges from the traditional, household unit of production, closely embedded in the local community, to the industrial, large-scale and globally operating corporations comprising the whole chain from harvesting to marketing. Governance must relate to powerful interests that are frequently in conflict over the scarce resource base that all participants share. Evading the cruel choices associated with zero-sum games, governments often favour a 'catch-all' policy. Such policies seldom 'add up', and the resource must then pay the difference. Sustainable resource use requires more than restrictions on harvesting practices and extraction volumes. It also infers choices pertaining the socio-economic structure of the industry as a whole. Should it be predominantly small-scale, labour-intensive, community-based, and following the model of flexible specialisation? (Piore and Sabel, 1984). Or should it be structured on Fordist principles: large- scale, high-tech, and capital intensive (Apostle *et al.*, 1998)? Few governments seem willing to take on the political burden associated with such a choice.

Industries have often-distinct institutional characteristics that make them different from each other. These differences are expressed in particular market connections, intra- and interorganisational relations, cognitive structures, and business cultures. Such patterns may also vary considerably from country to country, even for the same industry. This is also true for fisheries (Hannesson, 1996; Holden, 1997; Arnasson and Felt, 1995; Apostle *et al.*, 1998). It is safe to conclude that fisheries are among the most diverse of industries. It holds every conceivable problem and challenge of governance. Therefore, it is well suited for comparative analyses of socio-political governance and learning.

In Whitley's (1994) conceptualisation, markets may be analysed on the basis of three key questions:

➤   Which economic activities and resources are co-ordinated and controlled through private authority structures?
➤   How are market relations between economic actors organised and structured?
➤   What are the dominant ways of organising and controlling activities and resources within authority structures?

In this perspective, industries form 'business systems' which are 'forms of economic organisation which combine contrasting responses to these questions in different institutional contexts' (Whitley, 1994: 155). The hypothesis is that the particular response to each of these questions has a decisive impact on the system's ability to learn how to exploit new opportunities arising in markets as a result of technological and institutional change. As Friis and Vedsmand (1996) argue, there is little exchange between the repository of offshore and onshore knowledge in fisheries. They hold that information frequently gets lost, wasted and distorted by existing institutional relations. A case in point is the auction system that is so common to European raw fish sales. The auction system leads to the industry's slow response to market change because information does not easily 'trickle down' from consumer to producer:

There is no direct contact between the buyer and fisherman that only allows few opportunities for exchanging information, most of which a buyer obtains through the classification system and from what he can actually see.....The anonymous character of the transaction means that there is no detailed product specification or guarantees concerning place, time, method or handling of catch. This information is lost when the fishermen release the fish on the quayside (pp. 9-10).

This works both ways. Since institutions such as the auctions system shield the fishermen from consumer market events, Friis and Vedsmand find, fishermen do not plan their fishing activities in response to international market information and quality standards. They only focus their activities on catching opportunities and management schemes. A similar argument has been put forward for Norwegian fisheries (Hallenstvedt, 1982). Here, independent export organisations monopolised market information, thus reducing 'market intelligence' (Cornish, 1995) of both harvesters and processors. Consequently, the government introduced new legislation to make the chain more penetrable to market information. This happened when the law permitted processors forward integration in the chain and to establish their own export business. Thus, to make the flow of information reach the producer, more direct links must be established between those that consume and those that harvest the fish. However, as Friis and Vedsmand point out, this does not necessarily require vertical integration. Chain participants may form information networks among themselves. Associations of fish processors may play a constructive role here, a role that they have largely neglected in the past.

The market is only one way in which economic actors co-ordinate their transactions (Campbell *et. al.*, 1991; Coase, 1937; Williamson, 1975). If contracts do not sufficiently serve the interests of the parties involved, for instance because risks are too high and information too uncertain, actors will seek other forms of governance, such as networks or hierarchies. Each form represents different modes of governance and inter-organisational relationships, where the degree of centralised control and individual autonomy varies from one form to the other. Again, the operating principles and structural relationships that characterise each governance system are expected to affect the way information is accumulated and shared throughout the system as a whole, and hence, also the extent of interactive learning. In general, autonomous actors co-operating in a network, should, hypothetically, be more motivated for interactive learning than actors that follow orders within hierarchies. First of all, as individual profit centres, autonomous actors are responsible for their own survival. In Fordist command-and-control structures, responsibility, learning and decision-making involve only the 'top' of the organisation.

An understanding of learning opportunities prevailing within fisheries must start from the analysis of fisheries as a 'business system' within which social interaction occurs and relationships of exchange exist and are built. We need to keep an eye on the fact that fisheries have a variety of subsystems, each with distinct features and dynamics. We must also search for those state initiatives that may enhance or inhibit interactive learning at the individual, organisation and the business system level. Since the forms of interaction vary between subsystems, there is hardly a single way of promoting interactive learning within fisheries. Each subsystem provides different opportunities for such learning. The choice of system model(s) is therefore integral to the enhancement of competence within the fishing industry as a whole.

Patterns of Change in Fisheries

Business systems are not static. They may change rapidly under the influence of external or internal crises, or more evolutionary as a consequence of long-term and more pervasive economic, social or cultural forces. 'The extent to which business systems are liable to change depends on how integrated and cohesive they are, and how closely interconnected business systems are with mutually reinforcing institutions' (Whitley, 1994: 176). The more integrated and cohesive, the more rapid the change, because change spurs momentum as effects spill over to other elements of the business system. The change may be like dominoes: one actor undermines the basis of others. Conversely it may be an exponential growth process because of symbiotic mechanisms where one actor supports the other. Systemic, interdependent relationships, so typical of fisheries, are very likely to produce one or the other effect (Jentoft and Wadel, 1984). Therefore, change in fisheries is gradual; it tends to be sudden and abrupt. It is either collapse or boom. Interactive learning may be exactly the factor that prevents the former and triggers the other.

Two other hypotheses related to change in business systems are of interest here (Hulsink 1996: 17-22). In the first instance business systems 'converge' to remain competitive on global markets. Here, learning takes place in a broader - at least regional - institutional setting, in which market conditions are the prime learning incentive. There is less room for intrasystem interaction as change, and the learning process on which it is based, is 'anonymous' and large-scale. In the 'divergence' hypothesis, national styles and government policies play a greater role. Here, learning and change within business systems follow more distinct patterns because there is more room for reflection on unique mixes of social, economic, political and cultural factors.

Scott distinguishes between theories that explain why institutions change from how they change. We are more interested in the 'how' question, which Scott terms a 'process approach' (Scott, 1995: 66-73). '(O)utcomes appear to require long sequences of events... other events arise through a branching process such that once a particular choice occurs, other possibilities are foreclosed. Such outcomes are regarded as 'path dependent'. As the process goes, and 'through a variety of self-reinforcing feedback mechanisms', the number of alternatives and opportunities is narrowed down (Scott, 1995: 65). How learning proceeds throughout the institutional process, as a cause and effect and as a force shaping the change process itself, is interesting from a research point of view. Are choice options broadened or narrowed down as a consequence of learning?

At first glance, one would expect the former, as learning is supposed to broaden our perspectives. However, as paradigms do to science (Kuhn, 1962), learning often has the opposite effect. Academic disciplines have their distinct perspectives, problem definitions and solutions. Increased knowledge of the biology of fish resources is an important step towards better management, but the over-fishing problem requires a broader knowledge base. Although fisheries management has been perceived as predominantly a biological concern until now, it also requires insights into the economics and sociology of fisheries. Advances in the science of marine biology therefore have not been sufficient to avoid over-fishing.

Thus, the learning process may start out in an open-minded, exploratory fashion. After that, the focus is on the results of previous actions and how to correct for failures. What begins as double-loop learning will over time develop into single-loop learning. As long as things function satisfactorily, ambitions are adjusted to experiences, and there is little incentive to do better. When new opportunities arise throughout the process, which would require a more substantial reorientation, they are simply ignored.

Then there is the problem of 'false learning' (March, 1976), i.e. inferences are drawn from superficial observations. Mistakenly identified causal relationships may bring fatal results for organisations and business systems. Fisheries as a resource-based industry is particularly vulnerable to false learning. The resource is fugitive and subject to variation beyond human observation, control and sometimes even comprehension. The resource crisis leading to the moratorium on cod fishing in Atlantic Canada is a good illustration here. Its reasons are still unclear and intensively debated among experts.

The process leading up to the Canadian moratorium decision (Finlayson, 1994) is interesting in this regard. The stable catches of offshore trawlers throughout the late 1980s and early 1990s made both the fishing industry and the scientific community insensitive to claims on inshore developments, until they finally realised the critical situation of the cod stock and the need for drastic measures. These measures created massive unemployment, especially in rural Newfoundland where they rely entirely on the cod fishery. In this context we can appreciate Powell's observation that 'once things are institutionalised they tend to be relatively inert, that is resist efforts to change' (Powell and DiMaggio, 1991). In Canada, routine science was unable to detect what was about to happen. Part of the problem was the resistance to learn from new and unfamiliar sources. In Newfoundland, inshore fishermen had warned for years against what they observed on their local fishing grounds.

According to Powell, those actors that are the least subject to conformist pressures most often generate innovations. These actors are more likely to be at the periphery of the organisation or business system, or on the outside (Powell and DiMaggio, 1991: 197-198). Again, this general observation begs for research on fisheries systems. From where in the system have changes been initiated in the past, from within or from the environment? If Powell's thesis also holds true for fisheries, what could possibly make the actors at the core of the industry more entrepreneurial? Unless these actors become change-oriented, the business system will likely move ahead slowly and new opportunities remain neglected. Is is interesting to research what makes innovations spread from the periphery to the core of the entire business system.

Powell's thesis also calls for studies of entrepreneurial careers. The thesis suggests that entrepreneurs are often 'outsiders'. They have nontraditional backgrounds and knowledge. Nevertheless, as role models for other actors they create waves of learning throughout the industry. Thus, they open new avenues for innovation and change. Jentoft (1993) has argued that the success of salmon aquaculture in Norway can largely be explained from such a process. Norway's natural conditions, such as mild water temperatures and sheltered coasts, are conducive to aquaculture development. However, the fact that this new production form spread along the entire coast over a relatively short time span, can only be explained by the way pioneers made their social networks into arenas for interactive learning. Without this, salmon aquaculture could well have become a lost opportunity in Norway. Here the local community 'accidentally' provided the institutional setting for interactive learning. Without ignoring the effect of the geographical community, the potentials for interactive learning within 'virtual communities' need to be explored within a European context. It is particularly important in this day and age when information technology reduces physical distance as a hindrance to effective communication. The Internet has created entirely new opportunities for interactive learning. How the Internet is now changing communication relationships within fisheries is a challenge for social research.

In a study of complex and uncertain public decision-making processes, Berting (1996) stresses the importance of considering different rationality types, including Weber's (1978: 25ff) distinction between value and substantive rationality. As Van der Schans (1996) has demonstrated in a study of salmon aquaculture in Britain, adding Habermas' (1984) notion of communicative rationality yields an interesting set of standards for evaluating the theories and practices of fisheries governance. The Weberian distinction suggests that we should not only judge patterns and institutions in fisheries in a means-ends perspective but that they can be important in themselves. For instance, the cultures of fisheries cannot always be regarded from a utilitarian perspective. Neither can they simply be discarded for the sake of modernisation and profit maximisation without eroding the viability of coastal communities.

From a Habermasian perspective, interactive governance must be truly communicative, participants should support their proposals by making their images and world views explicit. It helps the communication process if each participant is able to see others' backgrounds, assumptions and aims. This also applies to interactive learning at the individual and system level. As Rist argues:

> Organisational learning takes place within the context of shared understandings, experiences, routines, values and acceptable behaviours. Organisational learning does not take place among complete strangers. ....Informal contact and a strategy of 'no surprises' establishes the level of interpersonal communication and trust necessary for the recipients of new information to incorporate that information into the organisation (Rist, 1994: 202).

From a communicative perspective, rational decisions must also be based on logical reasoning, empirically verifiable 'facts', controllable experiences and discussible interpretations. However, a discourse that takes place within a highly diverse, dynamic and complex socio-political context such as fisheries may easily go astray. Actors are likely to express different rationalities, as when small-scale fishers defend their livelihoods, corporations their profits, and environmentalists unspoilt nature. Chances are high that different insights and perspectives are deemed irrational, thus blocking the communicative process from continuing.

Perceptions of the rational approach to fisheries problems and opportunities are always related to social values and world views. Fisheries governance must balance natural, economic, social, and cultural concerns. This act is political rather than technical. Whatever notion of rationality is employed in governance, fisheries will need a surveillance system to detect and monitor performance on all these dimensions. Today such systems are mostly absent. Therefore, governance in fisheries tends to fail in *any* rationality test.

Scott differentiates between regulative, normative and cognitive 'pillars' of social institutions. This means that 'path-dependent' patterns of change can be examined for each of the three pillars. Without denying the importance of the other two, this Chapter is primarily concerned with the cognitive pillar and with potential changes in its elements in fisheries. Today, governments rely wholly on the regulative pillar, while the cognitive pillar is mostly overlooked. Thus, knowledge building as integral to fisheries management is an untapped opportunity in dealing with the problem of over-fishing. This argument also supports the advocacy for 'co-management' in fisheries (Jentoft, 1989). In fisheries management, second-order learning focusses on the regulative pillar as well. Where regulation fails, as it frequently does in fisheries, it leads to an often vicious circle of more and stricter regulations, which only creates more violations, which again call for even stricter regulations. This is yet another illustration of how the evolution of fisheries resource management is subject to 'path dependency'.

In strengthening the cognitive pillar of fisheries institutions, decision-makers may employ different strategies. One avenue is to look at innovation as a social-political system opportunity (Kooiman, 1997). The second avenue is more product or organisation-oriented. The two approaches can also support each other or become intertwined. In the social-political systems approach the diversity, dynamics and complexity of the opportunity must be 'handled' in terms of a cyclical multi-stage process. Initially we must look for the dynamics of knowledge with positive feedback potential. Following this, we need to identify participants from the perspective of the diversity of insights. The next step involves focusing on the relations involved in subsystems and interactions in their attempts to cope with complexity. The cycle may then be repeated (cf. for example Chapter 7, this volume).

From a product or organisation-oriented innovation perspective, the primary process in fisheries, the products the industry handles and the organisations involved receive the most specific focus. Here two forms of innovation scenarios can be discerned:

➢ a more institutionalised scenario emphasising semi-formalised interactions, in which market-oriented innovation dominates. The basis for this strategy is Research and Development and a diffusion-adoption strategy;
➢ a scenario based on the learning capacity of actors involved, and can be labelled as a learning and convincing strategy (Rogers and Shoemaker, 1971; Cozijnsen and Vrakking, 1986).

From this perspective it is apt to look into strategies and scenarios to be developed to exploit or enhance existing innovative capacities of fisheries systems, and the level of aggregation (organisations, sets of organisations or the industry as a whole) that would be most appropriate for developing and implementing innovative actions.

Inter-sectoral Learning

Insight into a subject area is not built from scratch, but draws from the body of knowledge, theories and concepts built up in other fields. Hence, knowledge is not a direct reflection of experience gained in the local setting. Rather, concepts and theories drawn from the general background mediate it. Nevertheless, the appreciation of the potential contribution of local experience and local knowledge for the improvement of fisheries management is growing (Ingles, 1993). Still, to validate this experience and knowledge to a larger non-local audience and to make it subject to interactive learning across local boundaries, it is essential that the local experience can be translated into concepts and language that this larger audience understands. Therefore, the process described above is not just a cognitive mechanism; it is also a social and in some ways even a political mechanism. A resource user must draw from his general background experience and from the reservoir of interpretations that his cultural background provides to structure his experience and to make it comprehensible to himself and to others. In trying to get his message across to (non-local) decision makers, the user must explain his local experience in the language and theories of these non-local decision makers. Hence, drawing from more general (more dominant) cultural reservoirs of interpretations to get the local experience across is also a strategic choice. It is a means to mobilise a larger political support base for one's ideas and aspirations.

This argument applies not only between fishermen and government agencies concerning ecological knowledge and fisheries management, but also to fisheries in relation to other sectors. Fisheries as a context of local experience has often been viewed from the perspective of agriculture and/or industry. Knowledge, concepts and methods developed in the agricultural and/or industrial setting are often used to understand (and manage) fisheries. In addition, to get their experience through on the political agenda, fishermen will often refer to the agricultural and/or industrial settings where things have been achieved that have not yet been achieved in fisheries. Lastly, the transfer of knowledge and concepts from agriculture to fisheries is facilitated by the fact that the Fisheries Department is a subsection of the Ministry of Agriculture in many European countries, and in general it is the smaller subsection at that. Policymakers or advisors with a background in agriculture will turn to fisheries at one point in their career. Later they move back into agriculture, the dominant sector in the Department. In the same way politicians specialised in agriculture also have fisheries in their portfolio, and researchers specialised in fisheries are employed by institutes that also research agriculture (and industry).

For instance, Holden argues that the EEC Common Agricultural Policy (CAP) was the blueprint for the Common Fisheries Policy (CFP). The CAP both aimed at making Europe self-subsistent and at preventing people in the countryside moving to the cities. Therefore, the EU poured money into agriculture to beef up production and to provide people with better incomes. By and large, the same policy was applied to fisheries. Thus, the CFP had a Structural Policy and a Market Policy long before there was a Conservation Policy. Consequently, European fleets expanded at rapid rates encouraged by available public money. In the Netherlands and elsewhere, fishermen warned that this was getting out of hand. Nevertheless, they were unable to push through a Horse Power Limit on grounds that this would distort competition.

It is another example of knowledge transfer with general validity in industrial production that does not always work out for fisheries that agreements between economic parties to limit production capacity are 'distorting' competition. Not until the introduction of the Conservation Policy were fisheries acknowledged to be different from agriculture, in that limits to growth differ between the sectors. Holden (1997) describes how knowledge transfer from agriculture to fisheries was simplified by an institutional link between the sectors at the bureaucratic and political level. Later, with the accession of new member states, fisheries obtained an independent Directorate General of its own. But then the damage, in terms of major fleet expansion, had already been done.

Analogies adopted from other settings help us to see fisheries problems in another light. 'What if we harvested juvenile sheep as recklessly as we do juvenile fish?' Most famous of all analogies employed in fisheries management is the Tragedy of the Commons, presented by Hardin (1968) in a seminal article on the overpopulation problem. Fish and fishermen may easily replace those cattle and herders on the commons that play a role in his parable, and we have a perfect model of the over-fishing problem. Thus analogies serve as metaphors and concepts do in scientific theory building. They help the ordering of individual cases and singular events into more general categories. However, knowledge transfer from one setting to another is not a simple metaphorical translation but a complex social and political process in which dominant frames of reference are often used to reinforce certain interests. Therefore, policies based on metaphors can be harmful. As Ostrom (1990: 23) argues, 'relying on metaphors as the foundation for policy advice can lead to results substantially different from those presumed to be likely.'

Neither is the contribution of the social sciences only the objective description and explanation of these processes. A mainstream body of knowledge in a scientific discipline is itself embedded in particular local, social, ecological settings and contains a diversity of mechanisms to interpret local experience. To transfer knowledge to other settings one must therefore acknowledge the differences between settings, the richness of local experience and the richness of science. The application of the Tragedy of the Commons model to fisheries has been criticised for violating this basic principle (McCay and Acheson, 1987).

Conclusion: Towards a Knowledge-Based Fishing Industry

This Chapter has argued that inter-organisational processes that involve and qualify the whole fisheries chain are key in making collective learning interactive. This is not to say that the European fishermen have been unable to learn in general. On the contrary, they have been very adaptive to modernisation. Technological developments in fisheries have been impressive, and fishermen have been quick to adjust to them. However, the advances have largely been concentrated on the catch and the adoption of ever more sophisticated gear and vessel technology. Today, however, the fishing industry has reached a saturation point. There are limitations on the degree of effectiveness with which the industry can utilise new forms of equipment without a qualitative addition to the knowledge base. Excellence in catching fish is not any longer sufficient in order to survive in global markets. When the fish is out of the water, the job has just begun, and it is not done until the product satisfies the consumer.

Important changes in fisheries are currently market-driven. The dynamics of the market globalisation, trade deregulation, increased international competition and powerful buying groups have been responsible for diversification or even polarisation of the market. These changes present both constraints and opportunities for the European fishing industry. These constraints will become even more apparent if the industry sticks to bulk production, with its low returns, lack of innovation and product development. A diversified market requires diversified fish products.

It is not necessarily a task for fishermen to market their fish, but fishermen need to know and understand the changing background for new demands on quality, environmental concerns and regulations. They also need to be able to exploit new ways of communication and the changing conditions for exchange. Large segments of European fishermen are not aware, and are not using, the growing knowledge base in these areas. Even in the short run, their incomes and survival depends on their ability to adapt to these new circumstances. Demands like these have made fishermen in many European countries upgrade their knowledge and seek government support for enhancing the level of competence in their industry. They also call for greater involvement and commitment from interest organisations such as fishermen's and fish workers' unions, where education was never high on the agenda.

The introduction of management systems at both national and international levels is another, and no less fundamental alteration of fishermen's working conditions. The institutions of fisheries management are no more than 15 to 20 years old, with the 200-mile EEZ in 1977 and the Common Fisheries Policy as milestones. This means that the same generation of fishermen has experienced a change from a predominantly open access and a free fishery, to a situation characterised by strict rules and limitations of their fishing practices. Adjusting to the new situation is a demanding change for fishermen and their interest organisations.

Altogether, management systems, market trends, and environmental concerns have radically transformed the policy area for the fishing industry. Most of all, they create an urgent need to expand the knowledge base among actors at every level of the industry. It is essential that fishermen, as primary producers, participate in an interactive learning process with fishermen's organisations and other players in the industry. This type of learning is quite different from the conventional 'learning-by-doing' that used to be so predominant in this industry.

Various institutional and organisational structures continue to produce path-dependency and to hinder the advance towards a knowledge-based industry. Market information is frequently lost, wasted or distorted by existing institutional relations. Producers in the mass market are unable to specify their requirements to fishermen as no formal mechanisms exist to allow such an exchange of information. Most of the infrastructure and institutions governing fishing activities are still geared on mass production. This mode of production assumes stability and control of the knowledge base as well. In the age of globalisation this assumption is unrealistic, and a continued reliance on it will inevitably lead to missed opportunities for growth and development in fisheries districts.

The continued survival of fishermen's organisations is questionable. Path-dependency within traditional functions, lack of financial and human resources, and weak external linkages reduce their ability to play any meaningful role in exchange of knowledge and information between the market and their membership. Most fishery organisations simply continue to carry out a function that individual market-intelligent fishermen could very well perform. There are only a few examples of local fishermen's organisations participating in promoting interactive learning within the industry. They frequently perceive themselves in a conflict and rights perspective, as defenders of a special interest, not actors that could take on responsibilities that would benefit the whole chain. Improving the conditions for their membership frequently requires initiatives that will also be in the collective interest, i.e. for the industry as a whole rather than segments of it. By pooling, organising and distributing information and knowledge, fishermen's, processors' and fish workers' organisations have the potential to function as an important interface between markets and their members.

Market changes will take place with or without the involvement of traditional fishermen and their organisations, with profound impacts on the structural relations within the industry. The chain is becoming more influenced by new institutions such as environmental organisations and global corporations within processing and retailing. These players have little patience in waiting for changes to be implemented, but create their own governance structures. To get a firmer grip on prices, the temperature chain and different quality levels the large retailers are now breaking new ways by circumnavigating the traditional supply chain. A case in point is Intermarché, the French retail chain. This actor is operating four stern trawlers to supply their supermarkets in France with fresh fish (Fishing News International, February 1997). Another example is the co-operation between Unilever, the food-processing giant, and the World Wildlife Fund, who have joined hands to form the Marine Stewardship Council to introduce to the consumer an eco-labeling scheme to prevent fish sale of over-fished stocks. Sainsbury, the British retail chain, has lent the new initiative its support from day one (ibid.).

The viability of the European fishing communities depends on their ability to meet the challenges from these global players. Traditional ways of doing business need to be reappraised and reformed, and new opportunities explored. Only a rapid, concerted and meaningful exchange of information and knowledge can achieve this. The volume orientation must be redirected towards the quality end of market. This would involve the whole chain, from harvesting to final sale. Not only does it infer a better performance by each actor, it also requires that their relationships, the ways they interact, be better co-ordinated. Our main argument in this Chapter follows by implication: learning must become interactive, a 'chain' process.

13 Creating Opportunities for Action

JAN KOOIMAN, MARTIJN VAN VLIET AND SVEIN JENTOFT

Introduction

This book has not so much looked at European fisheries from the perspective of the problems it is facing as from the perspective of opportunities. By focussing on opportunities rather than just problems, we were able to advance ideas that otherwise would not have been developed in a systematic manner. We argue that in an industry such as fisheries, the only way to devise a governance strategy that attempts to create opportunities is to *cross institutional boundaries*. By crossing institutional boundaries, opportunities for fisheries in Europe can be realised, not as *dei ex machina*, but as consciously and carefully handling a number of governing elements within the sector itself as well as in its immediate environment. First, this relates to *learning*. Governance must encourage interactive learning between all actors involved in the fisheries chain, i.e. from the fisherman to the consumers, and between government agencies at various levels of authority. Secondly, boundary-crossing relates to *responsibility*. All levels and forms of governance must be activated, not just the top level and not just the state.

Next, there is the *complexity* of fisheries. In principle public authorities are the institutions with most oversight and capacity to handle complex issues typical of a sector such as fisheries. The main lesson to be learned from theory and practical experience is that complexity reduction cannot be left to chance. Neither can it be done in an arbitrary manner. There are rules according to which complexity can be reduced to keep the loss of information, the effects of 'leaving things out', as low as possible. Although public authorities have a key role to play in dealing with the complexity aspects of fisheries issues and structures, they should not do this on their own. Governments rarely have sufficient knowledge about all (potential) aspects and their interrelations. Here coalitions between public and civil society partners (such as universities and other research institutions) will be a contribution. Interaction between public authorities and those in the other two societal institutions are badly needed, at least with respect to information exchange. Fisheries governance should be organised according to the principle of subsidiarity which, despite frequent lip service, is far from the case today. Central authorities should not carry out all functions of fisheries management, for reasons of democracy and equity, but also for reasons of efficiency. There is a strong case for devolving management responsibilities to lower levels of administration and to user-organisations. This is particularly true in an industry that is characterised by great diversity, dynamics and complexity, which is certainly true of the fishing industry.

A governance model that has the attention of nearly all Chapters of this book is *co-management*, defined here as *sharing management responsibility between user groups, government agencies and research institutions*. In itself, this model is an expression of boundary crossing. It transfers management authority between institutions: from top-down to bottom-up. The fishing industry is a prime example of a modern socio-political sector in which traditional command-and-control modes of governing are inadequate and often counter-productive. Rather, more interactive modes of governance are needed to create new opportunities and to realise the institutional frameworks promoting such interactions. This model is more than an option: it is a necessity especially for an industry that emphasises the importance of learning - which the fishing industry must do to be sustainable in an increasingly global environment.

The book contains a mixture of conceptual and empirical contributions. Both are needed in order to make opportunity creation more than a slogan. There is a need to reconceptualise many of the current issues in fisheries governance. A convincing argument must be theoretically consistent and empirically supported.

The Governance Approach

The governance approach conceptualises the problems of fisheries different from traditional approaches. Often the problem of fisheries management is considered to be an economic problem in the harvesting sector caused by fishermen's behaviour. By contrast, the governance approach takes into account that the fishing industry forms an interactive socio-economic and ecological system embedded in institutions, social networks, and cultures. In this system processors, distributors, and consumers are important players as well. There would be no overexploitation of stocks if the fish were solely consumed by the fishermen and their families. The behaviour of fishermen depends on their relation to the other agents and actions in the whole system, including government agencies.

The governance approach acknowledges the need to take into account a broader set of concerns than economic ones. Such concerns are related to the effects of fishing on the ecological system, social concerns such as unemployment, the sustenance of fishery-dependent regions and coastal communities, political concerns related to the legitimacy vis-à-vis users, the general public, and neighbour-states, as well as the feasibility of policies.

The problems and failures of management are most clearly revealed at the beginning of the chain: disrupted ecosystems, declining catches, laid-off fishing fleets, and shut-down processing plants. However, there is no reason to think that solutions are always found at the same location where problems emerge. The governance perspective focuses on the social and economic interactions within the various sectors of fish supply chain. In contrast to top-down public policy approaches, the governance approach stresses that influencing fisheries is more than government regulating fishermen, seen as adaptive objects of government intervention. In contrast to economic approaches, it emphasises that 'getting the incentives right' is a far more diverse, dynamic and complex process than creating once-and-for-all financial incentives for the fishing sector. The perspective is sensitive to the effects of quality regulation and quality management in the food sector. It has an eye for the concentration tendencies in the food industry and retail, the control strategies of international food companies, health and environmentally aware consumer behaviour etc. on fisheries. Furthermore, the governance approach has a focus on the interactions between the chain and its environment. Opportunities for sustainable fisheries management should be looked for in every part of the chain and in its environment.

In diverse, dynamic and complex social areas such as fisheries, a single authority will be unable to realise effective governance. The complex, dynamic and diverse character of fisheries makes it impossible for the collective responsibility of managing fisheries to be pursued in a top-down structure in which 'the government', either national or European, is in full control. Governments have many instruments and tools for steering and controlling, but they (often) lack the necessary feeling for what is going on at the grassroots level and miss the orientation and capacity for fine-tuning. Lindblom's famous phrase applies here: 'governments have strong thumbs but no fingers' (Lindblom, 1977).

Effective governance is achieved by the creation of interactive, social-political structures and processes that stimulate communication between the actors involved and the creation of common and shared responsibilities. The question to be addressed is how this new mode of governance is best put into practice in contemporary fisheries in Europe and elsewhere. The interactive mode of governance seeks opportunities for involving various stakeholders in the decision-making and management process, such that not only continuous organisational and inter-organisational learning processes are promoted (see this volume, Chapter 12), but also co-ordinated action.

There is a need to restructure governing responsibilities, tasks and actions based upon the integration of various concerns and the agents representing them. In a world characterised by increasing diversity, dynamics and complexity, uncertainty is the rule, not the exception:

➢ Social-political problems are the result of various interacting factors that are rarely wholly known and that are not caused by a single factor.
➢ Technical and political knowledge about problems and possible solutions is dispersed over many actors.
➢ Public policy objectives are not easy to define and are often submitted to revision (Kooiman, 1993: 255).

A top-down perspective or a rational, centralised approach to governance and management has therefore become increasingly out of step with the tasks they seek to address. Management techniques and policy instruments based on this approach are becoming more and more 'implementation-intensive' and 'enforcement-expensive' (Dunsire, 1993: 24). This certainly applies to the fisheries sector and its management. One of the main problems with dominant modes of fisheries management is that they follow Garrett Hardin's message too literally and plead for privatisation or state intervention. They pay no attention to the co-operative mode of governance, which in theoretical as well as practical terms might be more advantageous (see Chapter 4).

In the European context, one needs to be aware of the difficulty of designing rules and procedures for gathering information and making regulatory decisions, of making actual decisions consented by EU as well as EEA members and implementing them. The paradox is that although everyone is aware that current decision procedures and current European rules are among the causes of problems in European fisheries (and, therefore, should be changed), there is less awareness of the fact that in order to change this situation a new perspective on the governing institutions of European fisheries is necessary (Chapters 4 and 7). What then, are the opportunities for the social-political governance perspective to be put in practice? Kooiman (1993: 251) has formulated conditions that are conducive to the emergence of social-political governance:

> Existing methods and instruments of governance are failing or subject to erosion;
> The issues to be solved are of great concern to the involved actors;
> Organisational forms and patterns of interest mediation are relatively new and not yet firmly established;
> Options for the creation of synergetic effects or win-win situations are present.

The first two conditions are clearly met in European fisheries: within the European administration (e.g. Holden, 1994), the academic community (e.g. Crean and Symes, 1996), environmental organisations (documents by Greenpeace and Seas at Risk), and among the fishermen themselves (any issue of any fisheries newspaper gives ample evidence of this). The objectives, procedures and outcomes of the prevailing fisheries governance system in a European context are heavily disputed. Whatever the differences between all these stakeholders, they agree that the fish stocks and the dependent fisheries and regions are in a critical situation. However, options for win-win outcomes seem to be available - an outcome in which larger fish stocks are less intensely exploited is beneficial to all. The third condition seems to be the most problematic. Although the Common Fisheries Policy is fairly recent (1983), the procedures of decision-making on the European as well as the national levels exhibit great 'resistance to change'. Each of the agents involved has the capacity of obstructing changes they do not want to see effected, but none has the capacity to innovate by crossing institutional boundaries. As a result there is a danger of ongoing impasse.

The challenge for anyone involved in the governance of European fisheries is to make the interactions within the fisheries arena productive. The challenge is to integrate the need for co-ordinated collective action with the need for individual and decentralised flexibility and governance capacity. Therefore, the following sections will conceptualise a number of factors that need to be taken into consideration for a perspective on the opportunities for action to succeed. They are:

> interactions between state, market and civil society in coping with the diversity, dynamics and complexity of fisheries;
> what this means in terms of 'governing tasks' to be carried out;
> what changes this suggests pertaining to the Common Fisheries Policy.

Interactions Between State, Market and Civil Society

Who should interact with whom in order to create opportunities in European fisheries, and what kinds of governing issues does this raise? Within the state domain, authorities at the local, regional, national and supra-national level deal with fisheries, environmental protection and local/regional economic development. Then there is the market domain. In European fisheries, POs and FOs are potential platforms for opportunity creation (see Chapter 4). Although their capacities differ widely, there is a 'strategic' potential here that could well be enhanced. In fact, these organisations seem to be the 'nucleus' (hub) around which sufficient opportunity-creating capacity might be built into the market sector. Finally, there is civil society. Here for instance the scientific community organised in research institutions and universities is a relevant actor. So are voluntary organisations such as environmental and consumer groups, community organisations and women's associations. All of these groups may be capable of taking opportunity initiatives. Thus, one might conclude that there is potential for opportunity creation in all three societal sectors within and around European fisheries. However, in actual fact few sources are active.

Based on experiences with fisheries governance in Europe, there is a growing awareness that 'doing things together' might be a more effective way of handling dilemmas within fisheries than 'doing them alone'. Striking the proper balance between the state, market and civil society is the most urgent and appropriate design task for such a management system. The mutual dependencies such as those between catch and market and between public and private spheres are demonstrated by several of the case studies in this book. Only by representing within the industry the characteristics of the external contexts and forces, and representing externally the main characteristics of the internal situations and conditions will there be a chance to develop appropriate answers to the challenges this industry is facing.

First, the *diversity* of fisheries is expressed in the different regions of Europe, such as Northern Spain, Scotland and the Netherlands. These regions have heir specific social, cultural and economic histories and present-day qualities which governance must relate to. It must also take into consideration the diversity among various types of fishery and aquaculture. Similarly, governing regimes and practices vary between EU member states, again with their historical and political-cultural and structural differences. There is no way such differences can be placed under one general authority. Rather, a balance has to be struck between governing needs and capacities of all those different types of fisheries. Representation of this diversity is a necessary governance condition (see Chapter 4).

Handling *dynamics* requires a different kind of governing response. From theoretical and experiential evidence, it is clear that hierarchical forms of governance, with their democratic safeguards, are not adequate in situations with a high degree of dynamics. Here, the market mixed with co-governance arrangements is the most effective governing mode. In coping with non-linear developments, shorter and more rapid feedback mechanisms are required. Here, market and civil-society partners are better equipped to organise such cycles than public authorities.

Finally, there is the *complexity* of fisheries. In principle public authorities are the institutions with most oversight and capacity to handle complex issues typical of a sector such as fisheries. The main lesson to be learned from theory and practical experience is that complexity reduction cannot be left to chance. Neither can it be done in an arbitrary manner. There are rules according to which complexity can be reduced to keep the loss of information, the effects of 'leaving things out', as low as possible. Although public authorities have a key role to play in dealing with the complexity aspects of fisheries issues and structures, they should not do this on their own. Governments rarely have sufficient knowledge about all (potential) aspects and their interrelations. Here coalitions between public and civil society partners (such as universities and other research institutions) will be a contribution. Interaction between public authorities and those in the other two societal institutions are badly needed, at least with respect to information exchange.

The argument can be summarised as follows:

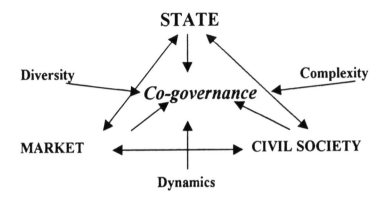

As the figure indicates, we argue that in order to deal with the diversity of fisheries, in particular 'co' forms between state and market are appropriate. In dealing with the dynamics of fisheries, particularly 'co' forms between market and civil society are in order. For the handling of complexity of fisheries, 'co' forms between public and civil society forms of governing seem to be the most effective mode of governance.

Governance Tasks in Diverse, Dynamic and Complex Fisheries

Governance of fisheries can be considered the pattern or structure that is the common outcome of efforts in intervention by all actors concerned, in the public as well as the private domain. The pattern is irreducible to the creation by one actor or group of actors in particular: 'political governance in modern societies can no longer be conceived in terms of external governmental control of society but emerges from a plurality of governing actors' (Marin and Mayntz, 1991: backflap). This governing order is a restricting, but also an enabling or reinforcing condition for social-political action. In practical action, such as creating an opportunity in a particular aspect of fisheries, there is a continuous interaction between restricting and enabling conditions, the use and the implementation of these conditions. A governance structure in fisheries has to release and co-ordinate sufficient transforming capacities in order to cope with all the aspects the creation of such an opportunity entails.

Contemporary governing of fisheries should primarily be seen as efforts to activate and co-ordinate social and economic actors, such that public and non-public interventions not only satisfy the needs of fishermen and their market partners, the communities they live and work in, but also the needs of those having a stake in the environmental and cultural aspects of fish, fishing, seas, rivers and oceans. Governing today's fisheries is predominantly a process of creating conditions such as co-ordinating public and private fisheries actors (Kauffman et al.,1986), balancing social and economic interests in fisheries (Dunsire, 1993) and creating and delimiting the scope (Mayntz, 1993) for social actors and systems in fisheries to organise themselves. Central here is the question how in practice mixes of different modes of governance (hierarchical, co-arrangements and self-governance) can be directed towards the enhancement of common fisheries goals, and how situations can be created where political governing and social organisation can be made complementary.

The argument so far has been that changes in socio-political sectors such as fisheries have brought out new governing needs, which have to be answered by new governing capacities. The sustainable development of fisheries requires new capacities that are more geared to 'doing things together', to co-operative interaction, than to everybody going it alone or centralised steering. What, then, are the specific governing tasks in fisheries? A classification of such tasks may look like this:

Coping with	*Tasks*
Diversity of fisheries:	integration and co-ordination
Dynamics of fisheries:	collibration and steering
Complexity of fisheries:	(de)-composition and regulation

'(De)composition' acknowledges the need to judge which social sectors and processes are involved and the relations between those involved, for every new problem or opportunity in fisheries. This requirement pertains to the issue of *complexity*. (De)composition depends mainly on the way a problem or an opportunity is defined: this definition process is basically a process of decomposing and compositing parts, processes and interaction patterns. Once a problem situation or an opportunity in fisheries has been defined the co-ordination 'game' will begin. The two tasks are closely interrelated and are basically complementary; in other words: there can be no co-ordination without (de)composition and vice versa. Here, the governance differs from the neo-corporatist approach in that in the latter, actors involved are precisely defined and co-ordination takes place within well-organised structures. In the governance approach this may vary: one situation will call for the presence of official representatives, in another situation the need for creativity requires other, less organised, and more independent actors. At least two ways of broadening the more traditional approach can be mentioned in this context. One way is to look at fisheries from the chain perspective, as we have been doing in Chapter 6 and 10. A second approach is to include environmental contexts, as in the case of aquaculture (Chapter 5). This can be considered as a case of 'composition'. Troadec (Chapter 9), for one, proposes situations of de-composition, distinguished different zones of the resource.

Steering and 'collibration' are ways to cope with the dynamics of modern fisheries. Steering is the definition of a norm and the means by which these norms can be reached. To 'collibrate' is to influence (social) power relations in such ways as to move social developments in a preferred direction (Dunsire, 1993). An example of this is the conscious inclusion of environmental interests in decisions on site licensing in fish farming. Steering is the more common form of governing, but tends to be 'implementation-intensive' and 'enforcement-expensive' (*ibid.*). Collibration is attractive for its relative simplicity: existing (political, administrative or social) processes can be guided without being forced to specify in advance what has to happen. It makes use of already existing (natural) feedback processes by amplifying or reducing them. Governing by positive or negative feedback is simpler than feed-forward in terms of the knowledge and information needed beforehand, which is one of the great difficulties in steering and other forms of feed-forward in coping with dynamics. Steering and collibration can be seen to complement each other: steering is about giving directions, while collibration is about indicating margins. Both belong together, and both are basic governing tasks.

Likewise, integration and regulation are strongly complementary tasks, directed at coping with the diversity of goals, demands, ideals and interests in modern societies. Integration points at the need for togetherness and shared meaning in the face of diversity. Regulation aims to overcome the externality problem, i.e. the negative effects of each individual (sub-system) actions on another, such as in environmental regulation: countering unwanted side effects of production and consumption. Integration works at a more fundamental level: in a well-integrated society actors can be more easily approached to account for the negative side effects of their behaviour.

The European Context

The context in which CFP revision is to take place has to be taken into consideration. This follows from our approach, which distinguish between governing in the sense of solving problems or creating opportunities (1st-order governing) and the institutional context in which this 1st-order governing takes place (2nd-order governing). To contribute to the debate on CFP revisions, then, we shall also have to develop ideas and suggestions on the way the EU as an institutional framework enhances or limits problem solving and/or opportunity creation. What follows are some key points that should be taken into account when revising the CFP.

A Multilevel Approach

The revision of the CFP does not take place in a void. The European Union is an ongoing policy-making process and legislative arena with many characteristics of its own. Although not as highly institutionalised and formalised as most governmental systems, it does not make sense to come up with proposals for revision which do not fit in with established patterns of EU practice. This applies to proposals for 1st-order (direct problem solving or opportunity creation) as well as 2nd-order governing, that is: institutional reform.

Notably, policy processes at the EU level are not limited to 'Brussels'. Community policy-making needs to be considered at four levels:

➤ the global, which is the least institutionalised but is becoming more and more the 'direct' or 'relevant' environment of almost everything taking place in the EU context;
➤ the supra-national level, which is the level of the European institutions in the narrow sense;
➤ the national or membership level;
➤ the sub-national levels.

As will have become clear from the foregoing Chapters, each of these four levels exerts some form of influence on fisheries policies, while different elements of the CFP have different impacts on the four levels.

Policymaking and Policy Implementation at the EU level

As Andersen and Eliassen point out:

> Formal institutions and legislation provide a framework for policy-making, but there is considerable room for different outcomes. All formal systems have some degree of freedom in this sense. However it seems that the loose structure [of the EU], combined with complexity and heterogeneity, provides more room for process in determining outcomes than in national, political systems. Actors bring to bear their national styles, strategies and tactics. (Andersen and Eliassen, 1993: 256)

According to these and other authors, the amount of room available for (new) initiatives depends greatly on the sector, the patterns of institutionalisation and EU history. There is a history of EU involvement in fisheries at the policy level. Yet, at the institutional level fisheries does seem to show a certain measure of openness for reform.

However, a clear distinction has to be made between policy preparation and policy implementation. The greater freedom in conceptualising new policies or revising existing ones is sharply offset by the actual implementation of policies. In this domain it is quite clear that the European Union, compared to the ability of national political systems of carrying out what it has decided, has much less leeway. It is largely up to the Member States to do what they are supposed to do in the EU context (subsidiarity). This is another reason not to frame suggestions for a new CFP without taking into consideration the relation between the EU and its member States.

Policymaking in the EU - Member State Relation

A development towards the completion of the internal EU market naturally reduces the capacity of member states to shape policies according to their own wishes. It can also be said that the EU's policy-making capacity cannot be increased to fully compensate for the loss of state control at the national level. This can be particularly problematic when taking into consideration the (normative) democratic aspect. As long as the European Union lacks a genuine democratic legitimation, one should be careful in extending EU power and reducing that of the national governments. The challenge is to increase the capacity of the European Union to improve the policymaking capacities at the EU level and at the same time to defend or regain the problem-solving capacities of the member states. Thus, the proposals for change should try to aim at 'win-win' outcomes. According to Scharpf (1994), this could be done on the basis of three principles:

➢ Different forms of co-ordination can be used to achieve similar purposes, while differing significantly in the degree to which they restrict the freedom of co-ordinated sub-systems;
➢ Regulation techniques are available (and tested in the EU context) that are less restrictive of national policy choices than the often practised strategy of harmonisation;
➢ Member states have to adopt policies that are compatible with EU objectives. In the years to come, experiments in fisheries can be set up according to certain principles.

Role of Subnational Levels

The future of subnational presence in Europe depends greatly on one's conception of the development of the EU as such. Will it be in the direction of a basically state-centred, supra-national, or multi-level European Union? The role of sub-national representation at the EU level differs considerably between three 'models' of the Union's future development. Roughly speaking, its influence will remain limited in the state-centred mode. In a supra-national model the state level may be eroded and sub-national units may be supported by EU institutions. In a multi-level model regions may become a governmental level of their own, and function in terms of networks next to national levels in relation to 'Brussels' (cf. the representation of German *Laender*). In proposals for a more formalised role and more substantive participation of regions or sub-national participation in the formation and execution of EU fisheries policies such (potential) developments must be taken into consideration.

In sum, any proposal for the revision of the CFP has to include the following aspects:

➢ consequences for each for the four relevant institutional levels;
➢ exploration of the degrees of freedom available or to be created in terms of policy space and institutional reform;
➢ 'win-win' governance outcomes in the relation between the EU, the national and sub-national levels.

Fisheries - a Special Case?

This book started from the conviction that fisheries should be analysed from a broad governance perspective, as this would force us not to focus solely on the problems that this industry is ridden with. It would allow us to examine the opportunities for making this industry a viable livelihood of the thousands and thousands of people throughout Europe that depend on it. For governance to be redesigned, it must first be creatively rethought at a conceptual level. Hence the theme and title of this book: *Creative Governance: Opportunities for Fisheries in Europe*. In doing so, it becomes clear that the issues confronting fisheries are of a very general nature, affecting the capacity of governance to address challenges that affect many sectors and industries. Just like other industries based on the utilisation of natural resources, the fishing industry must learn to live with a very complex, dynamic and unpredictable environment. This creates a demand for modes of governance that allow the industry to be flexible and adaptive. The fishery must be able to perform effectively in an increasingly global environment. This, however, is not only true of fisheries, but an increasing number of other industries as well. They all now find themselves to be inhabitants of the 'global village', and must learn to act accordingly. For this they should not have to be paralysed by their problems, but rather become inspired by their opportunities. A key resource here is competence, which again requires the ability to learn.

One consequence of economic globalisation is the standardising and universalising impact on industries, communities, and cultures. It is precisely these diversities that provide competitive advantages and promote interactive learning. Just as there are strengths in bio-diversity, cultural diversity is a positive force. Thus, in the face of globalisation, diversity as a resource risks being decimated, similar to our natural resource base (biodiversity).

Sustainable development has a natural as well as a socio-cultural basis. This holds true for fisheries as well as other industries. Therefore, the case of fisheries has something to offer to the learning process that is now going on in society as a whole. The problems and opportunities of this industry are very similar, if not identical, to those affecting other natural resource-based industries. Although fisheries in Europe have their particular characteristics which set the industry apart in many ways, we also believe that the theoretical approach and the conceptual issues raised in this book can be fruitfully applied to other resource-based industries with comparable diversity, dynamics and complexity. The fishery is a rich domain - many of its characteristics go far back in history. It is also a sector in which modernisation penetrates at a rapid pace and where these traditional (harvesting and local or regional marketing) aspects are in stark confrontation with modern technological and economic developments. Thus, the governance of fisheries may serve as a model for other traditional industries facing modernisation and globalisation.

Our model of co-governance is sufficiently general to enable application to other European industries. The debate on the specific functions of societal institutions (state, market and civil society) and the opportunities that may be created as a consequence of their closer interactions and the crossing of institutional boundaries, is relevant to situations outside fisheries. Therefore, rethinking fisheries governance is a step in building models for the broader institutional reform of governance.

References

Acheson, J.M. (1981), 'Anthropology of Fishing', *Annual Review of Anthropology* 10: 275-316.

Alegret, J.L. (1996), *Property rights, regulatory measures and the strategic response of Catalan fishermen.* Paper presented at the Workshop of the European Social Science Fisheries Network, 5-7 September, Sevilla.

Allen, P. and McGlade, J.M. (1987), 'Modelling complex human systems: a fisheries example', *European. Journal of Operations Research* 20: 147-167.

Andersen, S.S. and Eliassen, K.A. (Eds.) (1993), *Making policy in Europe.* Sage, London.

Apostle, R., Barrett, G., Holm, P., Jentoft, S., Mazany, L., Mikalsen, K.H. and McCay, B. (1998), *Community, state, and market on the North-Atlantic Rim. Sustainable fisheries in Atlantic Canada and North Norway.* Toronto University Press, Toronto.

Argyris, C. (1992), *On organizational learning.* Blackwell, Cambridge.

Arnason, R. (1996), 'On the ITQ fisheries management system in Iceland'. *Reviews in Fish Biology and Fisheries* 6 (1): 63-90.

Arnasson, R. and Felt, L. (Eds.) (1995) *The North Atlantic fisheries: successes, failures & callenges.* Island Living Series, the Institute of Island Studies, Charlottetown (P.E.I).

Arrow, K.J. (1984), *The Economics of Information.* Harvard University Press, Cambridge (Mass.).

Balton, D.A. (1996), 'Strengthening the Law of the Sea: The new agreement on straddling fish stocks and highly migratory fish stocks'. *Ocean Development & International Law* 27: 125-151.

Bardach, E. and Kagan, R.A. (1982), *Going by the book: the problem of regulating unreasonableness.* Temple University Press, Philadelphia.

Bateson, G. (1972), *Steps to an ecology of mind.* Ballantine Books, New York.

Baumol, W.J. and Oates, W.E. (1988), *The theory of environmental policy.* (2nd ed.) Cambridge University Press, Cambridge.

Bennett, J.W. (1976), *The ecological transition. Cultural anthropology and human adaptation.* Pergamon Press, New York.

Berkes, F. (Ed.) (1989), *Common property resources, ecology and community-based sustainable development.* Belhaven Press, London.

Berting, J. (1996), 'Over rationaliteit en complexiteit' in Nijkamp, P. *et al.* (Eds.) *Denken over complexe besluitvorming*. SDU, 's Gravenhage, 17-30.

Biesheuvel (1992), *Rapport van de Stuurgroep Biesheuvel, Beheerst vissen*. Visserijcentrum, Rijswijk.

Bijman, J., Tulder, R. and van Vliet, M. (1997), *Agribusiness, R&D en internationalisatie. Internationaliseringsstrategieën van agro-ondernemingen en de betekenis voor het eigen kennismanagement.* NRLO-rapport 97/12, NRLO, The Hague.

Boulding, K. (1968), *Beyond economics. Essays on society, religion and ethics.* University of Michigan Press, Ann Arbor.

Campell, J.L. *et al.*(Eds.) (1991), *Governance of the American economy.* Cambridge University Press, Cambridge.

Christensen, V. (1992), 'Managing fisheries involving predator and prey species'. *Review of Fish Biology and Fisheries* 6: 417 – 442.

Christensen V. and Pauly D. (1992), 'ECOPATH II - A software for balancing steady-state models and calculating network characteristics'. *Ecological Modelling* 6-1: 169 – 185.

Clark, E. (1964), *The Oysters of Locmariaquer.* The University of Chicago Press , Chicago.

Coase, R. (1937), 'The Nature of the Firm'. *Economica* 386-405.

Coase, R. (1960), 'The problem of social cost'. *Journal of Law and Economics* 3:1-44.

Cook, R.M., Sinclair A. and Stefansson, G. (1997), 'Potential collapse of North Sea cod stocks'. *Nature* 385: 521- 522.

Cornish, S. (1995), 'Marketing matters: The function of markets and marketing in the growth of firms and industries'. *Progress in Human Geography* 19 (3): 317-337.

Cornish, S. (1997), 'Product innovation and spatial dynamics of market intelligence: Does proximity to markets matter?'. *Economic Geography* 73 (2).

Cozijnsen, A.J. and Vrakking, W.J. (1986), *Handboek voor strategisch innoveren.* Kluwer, Deventer.

Crean, K. and Symes, D. (Eds.) (1996), *Fisheries management in crisis.* Blackwell Science, Oxford.

Crown Estate (CEC) (1987), *Fish farming, guidelines on the siting and design of marine fish farms in Scotland.* Crown Estate, Edinburgh.

Crown Estate (CEC) (1988), *Guidance notes on environmental assesment of marine salmon farms.* Crown Estate, Edinburgh.

Cunningham, S., Dunn, M.R. and Whitmarsh, D. (1985), *Fisheries economics, an introduction*. Manshall, London.

Curtil, O. (1996), 'La coquille St-Jacques en rade de Brest et le droit', in *'L'Économie brestoise 1995-96'*. Université de Bretagne Occidentale, CES-CEDEM, 107-131.

Davidse, W. P. and Beijert, G. (1995), *Mogelijkheden om investeringen in tong- en scholquota terug te verdienen*. LEI-DLO, The Hague.

Doel, C. (1996), 'Market development and organizational change: the case of the food industry' in N. Wrigley and M. Lowe, M. (Eds.), *Retailing,, consumption and capital: towards the new retail geography*. Longman, London, 48-68.

Downing, P.B. and Hanf, K.I. (Eds.) (1983), *International comparisons in implementing pollution laws*. Kluwer-Nijhoff, Boston.

Dryzek, J.S. (1987), *Rational ecology. Environment and political economy*. Basil Blackwell, London.

Dryzek, J.S. (1992), 'How far is it from Virginia and Rochester to Frankfurt?' *British Journal of Political Science* 22: 397-417.

Dubbink, W., van der Schans, J.W. and van Vliet, M. (1994), *'Vissen is veel meer een berekening geworden'*. (A report commissioned by the Dutch Ministry of Agriculture, Nature Management and Fisheries). Rotterdam School of Management, Dept. of Public Management, Erasmus University, Rotterdam.

Dubbink, W., van der Schans, J.W. and van Vliet, M. (1995), *Fishing has become more of a calculation but not all my colleagues are aware of it yet - the Dutch flat fish chain in transition*. Management Report Series, no 235, Rotterdam School of Management, Erasmus University Rotterdam, Rotterdam.

Dubbink, W. and van Vliet, M. (1995), *From ITQs to co-management: comparing the* usefulness of markets and co-management illustrated by the Dutch flatfish sector. Paper presented to the 10th Session of the ad hoc Expert Group on Fisheries. OECD, 3-6 October, Paris.

Dubbink, W. and van Vliet, M. (1996), 'Market regulation versus co-management?' *Marine Policy* 20 (6): 499-516.

Dubbink, W. and van Vliet, M. (1997), 'The Netherlands: from ITQ to co-management. Comparing the usefulness of markets and co-management illustrated by the Dutch flat fish sector', in OECD, *Towards Sustainable Fisheries: Issue Papers*. OECD, Paris, 177-202.

Dunsire, A. (1993), 'Modes of governance' in J.Kooiman (Ed.) *Modern governance*. Sage, London, 21-34.

Durrenberger, P. and Pálsson, G. (1985), 'Peasants, entrepreneurs and companies. The evolution of Icelandic fishing'. *Ethnos* 50 (1/2): 103-122.

Dijkema, R. (1988), 'Shellfish cultivation and fishery before and after a major flood barrier construction project in the Southwestern Netherlands'. *Journal of Shellfish Research* 7 (2): 241-252.

EC [European Commission] (1998), 'Report of the Group of Independent Experts to Advise the European Commission on the Fourth Generation of Multi-annual Guidance Programmes'. Report DG XIV/298/96.EN.

Eggertssen, T. (1990), *Economic behavior of institutions*. Cambridge University Press, Cambridge.

FAO (1992), 'Marine fisheries and the Law of the Sea: A decade of change'. *FAO Fish. Circ.*, 853, Rome.

FAO (1993), *The State of Food and Agriculture 1993*, FAO Agricultural Series No. 26, Rome.

FAO (1995), *State of the World Fisheries and Aquaculture*, FAO, Rome.

Feeny, D., Hanna, S. and McEvoy, F.A. (1996) 'Questioning the assumptions of the "Tragedy of the Commons" model of fisheries'. *Land Economics* 72 (2): 187-205.

Field (1987), 'The Evolution of Property Rights', *KYKLOS* 42(3): 319-345.

Finlayson, A.C. (1994), *Fishing for truth, a sociological analysis of Northern cod stock assessments from 1977 - 1990*. ISER, Newfoundland.

Flynn, A., Marsden, T. and Ward, N. (1994), 'Retailing, the food system and the regulatory state: constructing the consumer interest', in Lowe, Marsden and Whatmore (Eds.), *Regulating Agriculture: Critical perspectives on rural change* Vol. 5. Fulton, London.

Forey, D., Lundvall, B-Å. (1996), 'The knowledge-Based economy: From the economics of knowledge to the learning economy' in *Employment and growth in the knowledge-based economy*. OECD Documents, Paris.

Friis, P. (1996), 'The European Fishing Industry: Deregulation and the Market' in Crean, K., Symes D., (Eds.), *Fisheries Management in Crisis*. Blackwell Science, Oxford.

Friis, P and Vedsmand, T. (1996), *Can European Fishermen's Organisations respond to the dynamic changes taking place in the international market*. Paper presented at the Colloquium 'The Politics of Fishing', 17-18 September, Newcastle-Upon-Tyne.

Garcia, S.M. (1989), 'La recherche halieutique et l'aménagement: grandeur et servitude d'une symbiose' in J-P. Troadec (sous la dir.) *L'homme et les resources renouvelables. Essai sur l'aménagement d'une ressource commune renouvelable*, IFREMER, Paris, 711-743.

Garcia, S.M. and Newton, C. (1996), 'Current situation, trends, and prospects in world capture fisheries' in E.K. Pikitch, D.D. Huppert, and M.P. Sissenwine (Eds.), 'Global Trends: Fisheries Management', American Fisheries Society Symposium, 20. Bethesda. Maryland, USA, 3-27.

van Ginkel, R. (1988), Limited entry: panacea or palliative? Oystermen, state intervention and resource management in a Dutch maritime community. *Journal of Shellfish Research* 7(2): 309-317.

van Ginkel, R. (1989), '"Plunderers" into planters: Zeeland oystermen and the enclosure of the marine commons' in J. Boissevain and J. Verrips (Eds.) *Dutch dilemmas: anthropologists look at the Netherlands*. Van Gorcum, Assen, 89-105.

van Ginkel, R. (1990), 'Farming the edge of the sea: the sustainable development of Dutch mussel fishery'. *Maritime Anthropological Studies* 3(2): 49-67.

van Ginkel, R. (1991a), 'The musselmen of Yerseke: an ethnohistorical perspective'. in J.R. Durand, J. Lemoalle, J. Weber (Eds.), *Research and small-scale fisheries/La recherche face à la pêche artisanale*. 2 Vols. Editions de l'ORSTOM, Paris, 491-499.

van Ginkel, R. (1991b), 'The sea of bitterness: political process and ideology in a Dutch maritime community'. *Man N.S.* 26(4): 691-707.

van Ginkel, R. (1994a), 'One drop of luck weighs more than a bucketful of wisdom'. Success and the idiom and ideology of Dutch shellfish planters. *Ethnologia Europaea* 24(2): 155-166.

van Ginkel, R. (1994b), Tacking between Scylla and Charybdis. The adaptive dynamics of Texelian fishermen. *International Journal of Maritime History* 6(1): 215-229.

van Ginkel, R. (1995), Fishy resources and resourceful fishers. The marine commons and the adaptive strategies of Texel fishermen. *Netherlands Journal of Social Sciences* 31(1): 50-63.

van Ginkel, R. (1996), 'The abundant sea and her fates. Texelian oystermen and the marine commons, 1700 to 1932'. *Comparative Studies in Society and History* 38(2): 218-242.

Gislason, H. (1994), 'Ecosystem effects of fishing activities in the North Sea', *Marine Pollution Bulletin*, 29: 520-527.

Glasbergen, P. (Ed.) (1989), *Milieubeleid, theorie en praktijk*, VUGA, Amsterdam.

Goodin, R. E. (1986), 'The principle of Voluntary Agreement'. *Public Administration*, 64: 435-444.

Gordon, H.S. (1954), 'The economic theory of a common property resource: the fishery', *Journal of Political Economy* 62: 124-142.

Grunert, K.G., Baadsgaard, A., Larsen, H.H. and Madsen, T.K. (1996), *Market orientation in food and agriculture*. Kluwer Academic Publishers, Boston.

Habermas, J. (1984), *The theory of communicative action*, Vol. 1, Beacon Press, Boston.

Habermas, J. (1987), *The theory of communicative action*, Vol. 2, Beacon Press, Boston.

Hagler, M. (1995), 'Deforestation of the deep fishing and the state of the oceans'. *The Ecologist* 25(2/3), 74-79.

Hallentvedt, A. (1982), *Med lov og organisasjon*. Universitetsforlaget, Tromsø.

Hamilton, L.C. (1997), 'Management, adaptation and large-scale environmental change' in D. Symes (Ed.), *Property rights and regulatory systems in fisheries*. Fishing News Books, Oxford.

Hannesson, R. (1996), *Fisheries mismanagement: the case of the North Atlantic cod*. Fishing News Books, Oxford.

Hardin, G. (1968), 'The Tragedy of the Commons'. *Science* 162: 1243-1248.

Harriss, J., Hunter, J. and Lewis, C. (1997), *The New Institutional Economics and the Third World*, Routledge.

Hatcher, A. and Cunningham, S. (1994), *Phase I comparative report. The role of producers' organisations in the national fisheries management systems of three EU member states: the United Kingdom, France and the Netherlands*, CEMARE, University of Portsmouth, Portsmouth.

Healey, M.C. (1997), 'Comment: The interplay of policy, politics and science'. *Can. J. Fish. Aquat. Sci.* 54: 1427-1429.

Héral, M., Bacher, C. and Deslous-Paoli, J.M. (1989), 'La capacité biotique des bassins ostréicoles'. in J-P. Troadec (sous la dir.) *L'homme et les ressources halieutiques. Essai sur l'usage d'une ressource renouvelable*, IFREMER, Paris, 225-259.

Herfindahl, O., and Kneese, A.V. (1974), *Economic theory of natural resources*. Charles E. Merrill Publishing Company, Columbus (Ohio).

Hersoug, B. and Rånes, S. (1997), 'What is good for the fishermen, is good for the nation: co-management in the Norwegian fishing industry', *Ocean and Coastal Management*, 35 (2-3): 157-172.

von Hippel, E. (1988), *The Sources of Innovation.* Oxford University Press, Oxford.

Hodgson, G.M. (1988), *Economics and Institutions.* Polity Press, Cambridge.

Hoek, P.P.C. (1878), 'Oestercultuur in den vreemde en bij ons'. *Eigen Haard* 41: 389-392.

Holden, M. (1994), *The Common Fisheries Policy.* Fishing News Books, Oxford.

Holden, M. (1997), *The Common Fisheries Policy.* Fishing News Books, Oxford (2nd ed).

House of Commons Agriculture Committee (HCAC) (1990) *Fourth Report, Fish Farming in the UK,* Volume I and II. Her Majesty's Stationary Office, London.

Hulsink, W. (1996), *Do nations matter in a globalising industry?* Eburon, Delft.

Hutchings J.A., Walters, C. and Haedrich R.L. (1997a), 'Is scientific inquiry incompatible with government information control?', *Canadian Journal for Fisheries & Aquatic Science.* 54: 1198 – 1210.

Hutchings J.A., Walters, C. and Haedrich R.L. (1997b), 'Reply: Scientific inquiry and fish stock assessment in the Canadian Department of Fisheries and Oceans and Reply: The interplay of policy, politics, and science. *Canadian Journal for Fisheries & Aquatic Sciences* 54: 1430 –1431.

IFREMER-MAAF (1995), *Bio-geographical identification of English Channel fish and shellfish stocks.* IFREMER, RI DRV 93-028, Paris.

Inglis, J.T. (Ed) (1993), *Traditional ecological knowledge: concepts and cases.* International Development Research Centre, Ottawa.

Jentoft, S. (1989), 'Fisheries co-management, delegating government responsibility to fishermen's organisations'. *Marine Policy* (2): 137-54.

Jentoft, S. (1993), *'Dangling lines. The fisheries crisis and the future of coastal communities. The Norwegian experience,* ISER Books, St Johns.

Jentoft, S. (1994), *Decentralisation of fisheries management, a quest for subsidiarity.* Paper, VIth Annual Conference of the European Association of Fisheries Economists, March, Crete.

Jentoft, S. and McCay, B. (1995), 'User Participation in fisheries management, lessons drawn from international experiences'. *Marine Policy* 19 (3): 227-246.

Jentoft, S. and Wadel, C. (1984), *I samme bat: Lokale sysselsettingssystemer i fiskerinaeringen.* Universitetsforlaget, Oslo.

John, B. (1991), *Milford Haven waterway.* Pembrokeshire Coast National Park Area Guide. Greencroft Books. Cleglen Club, Cardiff.

Johnson, R.W.M., Petrey, L.A. and Schroter, W.R. (1996). 'Agribusiness: political economy and market structure. Toward a structure for agribusiness'. *Review of Marketing and Agricultural Economics*, Vol 64, No 2.

Kamien, M.I. and Schwarz, N.I. (1982), *Market structure and innovation.* Cambridge University Press, Cambridge.

Kaufmann, F.X. *et al.* (Eds.) (1986), *Guidance, control and evaluation in the public sector.* De Gruyter, New York.

Kearny A.T. (1994), *De markt gemist?*, Amsterdam: 2/16667.

Keesing, R.M. (1981), *Cultural anthropology. A contemporary perspective.* (2nd Ed). Holt, Rinehart and Winston, New York.

Kooiman, J. (Ed.) (1993), *Modern Governance: New government-society interactions.* Sage, London.

Kooiman, J. *et al.* (1997), *Social-political governance and management,* Management Report Series 34 (13), 3 Vols. Rotterdam School of Management, Section Public Management, Erasmus University, Rotterdam.

Kooiman, J. *et al.* (1998), *Co-governance of 'catch' and 'market'.* Final Report Contract EU FAIR Grant 94/c 185/08 DG XIV European Union.Rotterdam School of Management, Section Public Management, Erasmus University Rotterdam and Biological Sciences, University of Warwick, Coventry, UK, Rotterdam.

Kooiman, J. and van Vliet, M. (1995), 'Riding tandem: the case of co-governance'. *Demos* 7: 44-45.

Kristensen, P.H. and Sabel, C. (1997) ,'The small-holder economy in Denmark: the exception as variation'. in Sabel, C.F. and Zeitlin, J. (Eds.) *World of possibilities: flexibility and mass production in Western industrialization.* Cambridge University Press , Cambridge, 344-378.

Kuhn, T.S. (1962), *The structure of scientific revolutions.* University of Chicago Press, Chicago.

Lazonick, W. (1991), *Business Organization and the Myth of the Market Economy.* Cambridge: Cambridge University Press.

de Leeuw, A.C.J. (1986), *Organisaties: Management, analyse, ontwerp en verandering: een systeemvisie.* Van Gorcum, Assen.

Libecap, G.D. (1989), *Contracting for property rights.* Cambridge University Press, Cambridge.

Lindblom, C.E. (1977), *Politics and markets.* Basic Books, New York.

LNV (1995), *Tussen-evaluatie van de uitvoering van de voorstellen van de Stuurgroep Biesheuvel,* Ministerie van Landbouw, Visserij en Natuurbehoud, The Hague.

Löfgren, O. (1979), 'Marine ecotypes in preindustrial Sweden: A comparative discussion of Swedish peasant fishermen' in R. Andersen (Ed.), *North atlantic maritime cultures. Anthropological essays on changing adaptations.* Mouton, The Hague. 83-109.

Lundvall, B.-Å. (1988), 'Innovation as an interactive process: from user-producer interaction to the national system of innovation' in Dosi et al.(Ed.) *Technical change and economic theory.* Pinter, London.

Lundvall, B-Å. (1992), *National Systems of Innovation.* Pinter, London.

McCay, B. and Acheson, J.M. (Eds.) (1987) *The question of the commons: the culture and ecology of communal resources.* University of Arizona Press, Tucson.

McCay B. and Jentoft, S. (1996), 'From the Bottom Up: Participatory issues in fisheries', *Management, Society & Natural Resources* 9 (3): 237-250.

McCay, B. and Jentoft, S. (1996), 'Unvertrautes Gelaede: Gemeineigentum unter der Sozialwissenschaftlichen Lupe', *Kölner Zeitschrift für Soziologie und Sozialpsychologie.* Sonderheft 36, 272-292.

McGlade, J.M. (1989), 'Integrated Fisheries Management models: understanding the limits to marine resource exploitation' *American Fisheries Society Symposium* 6: 139- 165.

McGlade, J.M. and Hogarth, A. (1997), 'Sustainable management of coastal resources'. *Hydro* 1(5): 6 – 9.

McGlade, J.M. and Price, A.R.G. (1993), 'Mutidisciplinary modelling: an overview and practical implications for the governance of the Gulf region'. *Maine Pollution Bulletin* 27: 361-377.

McGlade, J. M. and Shepherd, J. (1992), 'Techniques for biological assessment in fisheries management'. *Berichte aus der Oekologischen Forschung* Vol. 9. TSBN 389336-091-3.

McKay, K.D. (1989), *A vision of greatness.* Maritime Museum, Milford, NH.

March, J.G. (1976), 'The technology of foolishness.' in J.G. March, J.P.Olsen *Ambiguity and choice in organizations.* Universitetsforlaget, Bergen.

March, J.G. and Olsen J.P. (1989), *Rediscovering institutions: the organizational basis of politics.* The Free Press, New York.

Marin, B. and Mayntz, R. (Eds.) (1991), *Policy networks.* Campus/Westview, Frankfurt/Boulder.

Marsden, T. (1996), 'Rural geography trend report: the social and political bases of rural restructuring', *Progress in Human Geography:* 20, 246-58. Arnold, London.

Marsden, T. and Wrigley, N. (1996), 'Retailing, the food system and the regulatory state' in N: Wrigley, N.and M. Lowe. (Eds.) *Retailing, Consumption and Capital: Towards the New Retail Geography.* Longman, London.

Marshall, A. (1920), *Principles of economics. An introductory volume* (8th edition) MacMillan, London.

Maus, I. (1986), *Rechtstheorie und Politische Theorie in Industrie-kapitalismus.* Wilhelm Fink Verlag, München.

Mayntz, R. *et al.* (1978). *Vollzugsprobleme der Umweltpolitik.* Kohlhammer, Wiesbaden.

Mayntz, R. (1993), 'Governing failures and the problem of governability' in J.Kooiman (Ed.), *Modern governance*, Sage, London, 9-20.

Ministerie van Landbouw, Natuurbeheer en Visserij (1993) *Vissen naar evenwicht, Structuurnota Zee- en kustvisserij,* SDU, The Hague.

Mintzberg, H. (1983), *Structures in five.* Prentice Hall, New Jersey.

Myers, R.A., Rosenberg, A.A., Mace, P.M, Barrowman, N. and Restrepo, V.R. (1994), 'In search of thresholds for recruitment overfishing', ICES, *Journal of Marine Science* 51: 191-205.

Nature Conservancy Council (NCC) (1989), *Fish Farming and the Safeguard of the Natural Marine Environment in Scotland.*

Neher, P.A. (1990), *Natural resource economics. Conservation and exploitation.* Cambridge University Press, Cambridge.

North, D. (1997), 'The new institutional economics and the third world'. in J. Harriss, J. Hunter. and C. Lewis, *The new institutional economics and the Third World.* Routledge, London.

North, D.C. and Thomas, P.R. (1973), *The rise of the Western world. A new economic history.* Cambridge Univ. Press, Cambridge.

Norton, G.A. (1984), *Resource Economics.* Edward Arnold, London.

OECD (1993), *Individual quota management,.* Paris.

Offe, C. (1984), 'Ungovernability; on the renaissance of conservative theories of crisis', in J. Habermas (Ed.), *Observations on 'The spiritual situation of the age'*. Harvard University Press, Cambridge (Mass).

Olson, M. (1965), *The logic of collective action*. Harvard University Press, Cambridge (Mass).

O'Riordan, B. (1996), *Tools for sustainable fisheries management*. Intermediate Technology Fisheries Case Studies.

Ostrom, E. (1990), *Governing the commons: The evolution of institutions for collective action*. Cambridge University Press, Cambridge.

Pearce, D.W. and Turner, R.K. (1990), *Economics of natural resources and the environment*. Harvester Wheatsheaf, New York.

Pearse, P.H. (1994), 'Fishing rights and fishing policy: the development of property rights as instruments of fisheries management' in *The state of the world's fisheries resources. Proceedings of the world fisheries congress planning session*, IBH Publishing Company, Oxford, 76-90.

Pearse, P.H. and Walters, C.J. (1992), 'Harvesting regulation under quota management systems for ocean fisheries'. *Marine Policy* 16 (3): 167-182.

Phillipson, J. and Crean, K. (1997), 'Alternative management systems for the UK fishing industry'. *Ocean and Coastal Management*, 35 (2-3): 185-200.

Pigou, A.C. (1920), *The economics of welfare* (Quotations from 4th edition). MacMillan, New York.

Piore, M.J. and Sabel, C. (1984), *The second industrial divide: Possibilities for prosperity*. Basic Books, New York.

Porter, M. (1980), *Competitive strategy*. Free Press, New York.

Powell, W.W. and DiMaggio, P.J. (1991), *The new institutionalism in organizational analysis*. Chicago University Press, Chicago.

Pressman, J.L. and Wildavski, A.B. (1973), *Implementation. How great expectations in Washingtion are dashed in Oakland* University of California Press, Berkeley.

Prigogine, I. and Stengers, I. (1988), *Order out of chaos*. Bantam, New York.

Raakjaer Nielsen, J., Vedsmand., T. and Friis, P. (1997), 'Danish fisheries co-management decision making and alternative management systems', *Ocean & Coastal Management* 35 (2-3): 201-216.

Rappaport, R.A. (1979), *Ecology, meaning, and religion*. North Atlantic Books, Berkeley.

Rhodes, R.A.W. (1997), *Understanding governance*, Open University Press, Buckingham.

Rist, R. C. (1994), 'The preconditions for learning: Lessons from the public sector' in F.L. Leeuw, R.C. Rist and C. Sonnichsen (Eds.), *Can governments Learn?* Transaction Publishers, New Brunswick.

Rogers, W.M. and Shoemaker, F. (1971), *Communication of innovations.* Free Press, New York.

Rokkan, S. (1971), *Stat, nasjon og klasse*, Universitetsforlaget, Oslo.

Ruigrok, W. and van Tulder, R. (1995), *The logic of international restructuring.* Routledge, London/New York.

Runge, C.F. (1986), Common property and collective action in economic development. *World Development* 14(5): 623-635.

Sagdhal, B. (1992), 'Co-management: a common denominator for a variety of organizational forms'. *NF-Report* Nr.17/92-20, Nordland Research Institute, Bodø.

Sahlins, M. (1974), *Stone age economics.* Adline Publishing Company, Chicago.

Salz, P. (1991), *De Europese Atlantische visserij, structuur, economische situatie en beleid.* LEI-DLO, The Hague.

van der Schans, J.W. (1993), *Governing marine salmon farming in Scotland, Colonialisation of the coastal frontier?*, Management Report Series No. 148, Rotterdam School of Management, Erasmus University Rotterdam, Rotterdam.

van der Schans, J.W. (1996), 'Governing in the face of diversity: the development of the salmon farming industry in Scotland', in T. Elfring *et al., European research paradigms in business studies.* Handelshojskolens Forlag, Copenhagen.

van der Schans, J.W. (1996), 'Colonizing the coastal frontier: Governing marine salmon farming in Scotland', in C. Bailey, S. Jentoft and P. Sinclair (Eds.), *Aquacultural development: social dimensions of an emerging industry.* Westview Press, Boulder.

van der Schans, J.W. (1997), *Kwaliteitsbeheer in de visserijsector: een co-management perspectief.* Management Report Series, 23 (13) Rotterdam School of Management, Erasmus University Rotterdam, Rotterdam.

van der Schans, J.W. (forthcoming), *Governance of marine resources.* Rotterdam School of Management, Rotterdam (dissertation).

Scharpf, F. (1994), 'Community and autonomy: multilevel policy-making in the European Union', *Journal of European Public Policy*, 1: 219-242.

Schelling, T.C. (Ed.) (1983), *Incentives for Environmental Protection.* MIT Press, Cambridge, Mass.

Schlager, E. and Ostrom, E. (1992), 'Property-rights regimes and natural resources: a conceptual analysis'. *Land Economics* 68(3): 249-262.

Scott, A. (1993), 'Obstacles to fishery self-government'. *Marine Resource Economics* 8: 187-199.

Scott, W.R. (1995), *Institutions and organizations.* Sage, London.

Scottish Office Agriculture and Fisheries Department (SOAFD) (1991), *Report of the Annual Survey of Fish Farms for 1991.*

Scottish Office Environment Department (SOEnD) (1991), *Guidance on the location of marine farms,* Consultative Draft.

Scottish Wildlife and Countryside Link (SWCL) (1988), *Marine fishfarming in Scotland,* Scottish Wildlife and Countryside Link, Perth.

Scottish Wildlife and Countryside Link (SWCL) (1990), *Marine salmon farming in Scotland - A Review.* Scottish Wildlife and Countryside Link, Perth.

Seafish Industry Authority (1995), *Catching for the market,* Edinborough.

Selznick, P. (1992), *The moral commonwealth: social theory and the promise of community.* University of California Press, Berkeley.

Sen, S. and Raakjaer Nielsen, J. (1996), 'Fisheries co-management: a comparative analysis'. *Marine Policy* 20: 405-18.

Stokes, K., McGlade, J.M. and Law, R. (Eds.) (1993), *The exploitation of evolving resources.* Lecture Notes in Mathematics, 99, Springer-Verlag, Berlin.

Stone, C.D, (1975), *Where the law ends. The social control of corporate behaviour.* Harper and Row, New York.

Storper, M. and Salais, R. (1997), *Worlds of Production, the Action Frameworks of the Economy.* Harvard University Press, Cambridge (Mass.).

Suarez de Vivero, J.-L. and Frieyro de Lara, M. (1997), Regions at sea: the administrative region as a base for an alternative fisheries management system for Spain, Special Issue of *Ocean and Coastal Management:* Fisheries Management in the North Atlantic: National and Regional Perspectives.

Symes, D. (1995), 'The European Pond: who actually manages the fisheries'. *Ocean and Coastal Management,* 27 (1-2): 29-46.

Symes, D. et al. (1996), *Devolved and regional management systems for fisheries.* Final Report of EU Project AIR-2CT93-1392: DGXIV), Hull.

Symes, D. (1997) 'Regionalisation of the Common Fisheries Policy', paper at ESSFIN Workshop on Alternative management systems for fisheries, Brest.

Symes, D. (forthcoming), 'Towards 2002: subsidiarity and the regionalisation of the Common Fisheries Policy' in T. S. Gray (Ed.), *The Politics of Fishing*, Macmillan, London.

Symes, D. and Phillipson, J. (1996), *The imperative of institutional reform: alternative models and the UK fishing industry*, 229-244 in the Proceedings of the VIIth Annual Conference of the European Association of Fisheries Economists, 10-12 April 1995, CEMARE: University of Portsmouth, Portsmouth.

Taylor, L.J. (1983), *Dutchmen on the bay. The ethnohistory of a contractual community*. University of Pennsylvania Press, Philadelphia.

Teknologi-Rådet (1996), *Fremtidens fiskeri*. Teknologi-Rådets rapporter 1996/8. København.

Tisdell, C. (1993), *Environmental economics. Politics for environmental management and sustainable development,*. Edward Elgar, Aldershot.

Townsend, R. E. (1992), 'Bankable individual transferable quotas'. *Marine Policy*, 16 (5): 345-348.

Townsend, R. E. (1995), 'Transferable dynamic stock rights', *Marine Policy*, 19 (2): 153-158.

Townsend, R.E. and Wilson, J.A. (1987), 'An economic view of the Tragedy of the Commons'. in B. McCay, and J.M. Acheson (Eds.) *The Question of the Commons. The culture and economy of communal resources*. University of Arizona Press, Tucson, 311-326.

Troadec, J-P. (1994), 'Le nouvel enjeu de la pêche: l'ajustement des institutions aux nouvelles conditions de rareté des ressources'. *C. R. Acad. Agric. Fr.*, 80(3): 41-60.

Troadec J-P. (1996), 'Produire mieux en pêchant moins: la régulation de l'accès' in 'Pêches maritimes françaises: bilan et perspectives'. *Revue POUR*, 149/150: 89-102; Diffusion L'Harmattan, Paris.

Tweede Kamer (1987), *Rapport van de Commissie Visquoteringsregelingen*. Vergaderjaar 1986-1987, 19955, No. 2.

van Vliet, M. (1993), 'Environmental regulation of business: options and constraints for communicative governance',.in J. Kooiman (Ed.), *Modern governance*. Sage, London.

van Vliet, M. (1998), 'Fishing as a Collective Enterprise – ITQs and the Organisation of Fishermen' in D. Symes (ed.), *Property Rights aned Regulatory Systems in Fisheries*. Blackwell, Fishing Newsbooks, Oxford, 67-79.

Vedsmand, T., Friis, P. and Raakjaer Nielsen, P. (1995), *Analysis of opportunities and constraints for alternative fisheries co-management models in Danish fisheries*, Phase III National Report, Devolved and Regional Management Systems for Fisheries, North Atlantic Regional Studies, Roskilde University, Roskilde.

Vedsmand, T. (1998), *Management and industrial development of the fisheries, seen from an institutional perspective* (in Danish). Research Centre of Bornholm and Department of Geography, Roskilde University.

Walters, C. J. and P.H. Pearse (1996), 'Stock information requirements for quota management systems in commercial fisheries'. *Reviews in Fish Biology and Fisheries*, 6 (1): 21-42.

Weale, A. (1992), *The politics of pollution*. Manchester University Press, Manchester.

Weber, Max (1978), *Economy and society. An outline of interpretive sociology*. Bd 1. Bedminster Press, New York.

Whitley, R. (1994), 'Dominant forms of economic organization in market economies'. *Organization Studies*, 15/2: 153-181.

Wicksell, K. (1934), *Lectures on political economy; Volume 1.* (cited from reprint, 1977 by Augustus M. Kelly Publishers, London).

Wilde, J.W. de (1993), *'Capacity and effort limitations in the Netherlands'*. Paper presented EAFE conference, Brussels (LEI/DLO).

Williamson, O. (1975), *Markets and hierarchies*. Free Press, New York.

Wilson, J.A. (1982), 'The Economical management of multispecies fisheries'. *Land Economics*, 58: 417-434.

Wise, M. (1996), 'Regional concepts in the development of the Common Fisheries: the case of the Atlantic Arc' in Crean, K., Symes, D. (Eds.) (1996) *Fisheries management in crisis*, Fishing News Books, Oxford, 141-158.

Printed and bound by CPI Group (UK) Ltd, Croydon, CR0 4YY

22/10/2024

01777628-0007